This Dynamic Universe

This Dynamic Universe

Edited by Corona Trew and E. Lester Smith

THE THEOSOPHICAL PUBLISHING HOUSE
Wheaton, Ill. U.S.A./Madras, India/London, England

© *Copyright 1983 The Theosophical Publishing House.*
All rights reserved. No part of this book may be reproduced in any manner without written permission except for quotations embodied in articles or reviews. For additional information write to: The Theosophical Publishing House, 306 West Geneva Road, Wheaton, Illinois 60189. Published by The Theosophical Publishing House, a department of The Theosophical Society in America.

Library of Congress Cataloging in Publication Data

This dynamic universe.

Republication, with revisions, of three works originally published 1951-1960 by the Theosophical Publishing House, London: This dynamic universe, This ordered universe, and Man's expanding horizon.
 1. Theosophy—Addresses, essays, lectures.
 2. Cosmology—Addresses, essays, lectures.
 I. Smith, E. Lester (Ernest Lester), 1904-
 BP570.T47   1983   299'934   83-9202
 ISBN 0-8356-0232-X (pbk.)

Printed in the United States of America

CONTENTS

Foreword vii

PART I — THIS DYNAMIC UNIVERSE

1. What is Fohat? 3
   E. Lester Smith, D.Sc., F.R.S., and
   Corona Trew, Ph.D., D.Sc.

2. Modes of Manifestation 10
   E. L. Gardner

3. Fohat and Intuitional Research 18
   E. L. Gardner

4. Fohat 24
   Josephine Ransom

5. References to Fohat in *The Secret Doctrine* .. 30
   E. Lester Smith, D.Sc., F.R.S.

PART II — THIS ORDERED UNIVERSE

Preface 47
V. W. Slater, B.Sc., F.R.S.C., F.I. Chem. E.

6. Introductory Principles 51
   Corona Trew, Ph.D., D.Sc.

7. The Triple Basis of Law 54
   Corona Trew, Ph.D., D.Sc.

8	Universal Law in Nature, Periodicity, Evolution and Karma	66
	Corona Trew, Ph.D., D.Sc., and Dorothy Ashton, B.Sc.	
9	Universal Law in Natural Forms and Organisms	75
	Corona Trew, Ph.D., D.Sc., and Dorothy Ashton, B.Sc.	
10	Man, the Measure of All Things	85
	E. L. Gardner	
11	Beyond Law	98
	Neville Reed, A.M.I.E.E., A.M.I. Mech. E.	
12	Immediate Applications	104
	E. Lester Smith. D.Sc., F.R.S.	

PART III — MAN'S EXPANDING HORIZON

	Introduction	111
13	The Universe as an Entity	113
14	Involution and Evolution	119
15	The Formative Impulses in Evolution	128
16	Man and Society — A Unified View	141
17	Man's True Role in the Universe	149
18	Values Beyond	157
	Epilogue	164

FOREWORD

DURING THE 1950's the Science Group of the English Theosophical Research Center, meeting periodically in London, considered the nature of the Universe. We examined such concepts as energy, order and purpose, as revealed by ancient philosophies reformulated as Theosophy, and by modern science and philosophy. The outcome was a trio of small books, or Transactions as they were called, published by the Theosophical Publishing House, London, with the titles:

This Dynamic Universe: Essays in Fohat (1951)

This Ordered Universe: A Study in Universal Law (1953)

Man's Expanding Horizon: This Purposeful Universe (1960)

These books have long been out of print. The Theosophical Publishing House in America recently accepted the suggestion of the surviving authors that they should be republished together, with extensive revisions agreed between us.

In a sense they are the forerunners of a later book from the Science Group, *Intelligence Came First*, published by the Theosophical Publishing House in America in 1975. Hence the overall title for the republished trio: *This Dynamic Universe*. A few scientists have recently expressed similar ideas, notably J. E. Lovelock in his book, *Gaia: A new look at life on earth* (1979). He wrote of Gaia, the Earth Goddess, as a "complex entity" and of the hypothesis "that the biosphere is a self-regulating entity with the capacity to keep our planet healthy by controlling the chemical and physical environment." Sir Bernard Lovell in his book *In the Centre of Immensities* (1978) tentatively expressed similar views in respect of the whole universe: "It seems that the chances of the existence of man on Earth today, or of intelligent life anywhere in the Universe, *are* vanishingly small. Is the Universe as it is, because it was necessary for the existence of man?"

In the first two of our books, chapters were credited to the authors who drafted them; but considerable redrafting was done in every instance, following discussions at one or more meetings of the Science Group. The third book was regarded as a Group effort and no attributions were made beyond a list of contributors and editors.

The following list includes the participants in all three ventures:

Dorothy Ashton, B.Sc.
H. Tudor Edmunds, M.B., B.S., M.R.C.S., L.R.C.P.
Edward L. Gardner
Richard Groves, M.Sc.
Josephine Ransom
Neville Reed, A.M.I.E.E., A.M.I. Mech. E.
V. Wallace Slater, B.Sc., F.R.S.C., F.I.Chem E.
E. Lester Smith, D.Sc., F.R.S.
Corona Trew, Ph.D., D.Sc.

PART I

THIS DYNAMIC UNIVERSE

Essays on Fohat

"It is through Fohat that the ideas of the Universal Mind are impressed upon Matter."
>*The Secret Doctrine*, Vol. 1, page 150

"... Fohat ... becomes on Earth the Great Power ... "
>Vol. 3, page 76

"... Fohat, the constructive Force of Cosmic Electricity ... polarized himself into positive and negative electricity."
>Vol. 1, page 201

References are all to the Adyar Edition of *The Secret Doctrine* unless otherwise noted.

CHAPTER 1

WHAT IS FOHAT?

by E. Lester Smith and Corona Trew

1. '... in the Manifested Universe, there is "that" which links Spirit to Matter, Subject to Object. This something, at present unknown to Western speculation, is called by Occultists Fohat.'
   The Secret Doctrine,
   Proem: Third Edition: I, 44
   Adyar Edition: I, 81

2. 'Fohat is the dynamic energy of Cosmic Ideation ... Fohat, in its various manifestations, is the mysterious link between Mind and Matter, the animating principle electrifying every atom into life.'
   Proem: Third Edition: I, 44
   Adyar Edition: I, 81

3. 'Fohat, then, is the personified electric vital power ... the action of which resembles ... that of a living Force created by Will, in those phenomena where the seemingly subjective acts on the seemingly objective, and propels it to action.'
   Third Edition, I, 136
   Adyar Edition: I, 170

4. 'Fohat is one thing in the yet Unmanifested Universe, and another in the phenomenal and Cosmic World. In the latter, he is that occult, electric, vital power, which under the Will of the Creative Logos, unites and brings together all forms, giving them the first impulse, which in time becomes law ... in the Unmanifested Universe Fohat ... is simply that potential creative Power....'
   Third Edition: I, 134
   Adyar Edition: I, 169

5. 'It is the action of Fohat upon a compound or even upon a simple body that produces life.'
   Third Edition: I, 573, *Footnote*
   Adyar Edition, II, 250, *Footnote 3*

FOHAT, as explained in *The Secret Doctrine* of H. P. Blavatsky, is the universal energy which includes all the forces of nature. It is that force which makes the Universe 'go', the driving force or fundamental energy from which all known kinds of energy derive. But Fohat is not just blind chaotic power, or rather it is not so when within the comprehension of a human mind. For men it is seen as directed, intelligent power—power inspired by purpose. That purpose is Evolution in its widest connotation, and Fohat is the Life-principle that bridges the poles of pristine spirit and inert Matter. It thus appears as both the energising and the stabilising factor in manifestation at all levels. It is the creative force forming the material of any one level and it is the binding energy linking together the individual units of that material. Since it is a force working in all the three regions of the cosmos, spiritual, intellectual and physical, it forms the uniting link between one of these regions and another. Thus it is the link between 'mind and matter' and between Spirit and Mind.

(References 1, 2 and 3 above)

In the Secret Doctrine concept of the Cosmos, the whole manifested universe is the result of the interaction of Cosmic Ideation with Cosmic Substance, and Fohat or Cosmic Energy is born of that interaction. It represents the result in the field of manifestation of that interaction, and so corresponds in nature to the force which generates tension and the resultant tension itself. (See especially *The Secret Doctrine, Third Edition* I, 350; *Adyar Edition* II, 41.)

In Man, human relationships represent another manifestation of that tension or Cosmic Energy.

Fohat is therefore a force which acts in various ways and in various regions of the universe. A sweeping definition to cover such grandeur of scope must contain

What is Fohat?

within itself many paradoxes. When it is seen as *the* force behind all manifestation, the method of correspondences may enable us to detect the characteristic mode of behavior uniting all its separate manifestations and common to them all. When it is expanded and analysed Fohat will appear as a bewildering series of powers and attributes that seem almost without connection and certainly without unity. Fohat thus appears in the guise of both servant and master; it is described as the messenger of the Gods, the creative forces of Nature, but also as the overseer of the gods (Rupa Devas and Nature Spirits).

Fohat manifests as fierce impersonal destructive energy yet also is personified as the Creator, and as the Pervader of Creation, sustaining it perpetually 'from one twilight to the other, during seven eternities'.

Fundamentally, Fohat is power in a dynamic form and it represents the creative purpose flowing through the manifested cosmos. In the Glossary of Theosophical Terms we find it defined as dual in manifestation. Such duality, however, is rather that of the two halves of a sine-wave form, Fohat representing the complete wave-form, including both its positive and negative phases.

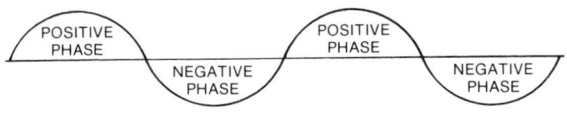

Diagram 1

It is said that 'Fohat' is the active (male) potency of the 'shakti' (female) reproductive power in nature, and 'shakti' is described as the 'active female energy of the Gods'. This would seem to imply that Fohat is the dynamic driving force thrusting down into manifestation and outwards at any one level, creating the vehicles and forms and binding them together, while shakti is the static receptive response of the material, isolating and maintaining the forms of the universe. To employ the analogy of a seal and wax, shakti is the capacity to take up an impression, inherent in wax and plastic materials, while Fohat is the potency that makes the impression, the seal, stamp or die itself, imprinting a replica of its own characteristic upon the receptive wax. Both Fohat and shakti represent differentiations of the larger universal Fohat as the One Force.

When manifesting at any one level Fohat appears in the two extreme modes of a distributive, scattering and expanding force, and a cohesive binding force. Electricity typically exemplifies its behavior. It is a force which can cause molecules to break down into elements. Yet, as it manifests within the atoms, the material forms it creates, it acts as a cohesive force holding together the oppositely charged fragments that compose the atomic form.

The present day scientific picture of the atom of matter is interesting when viewed in connection with the behaviour of Fohat. The atomic centre consists of a nucleus of energy comprised of positive units of electricity prevented from flying apart by the moderating effect of neutral particles or 'neutrons'. This positively charged centre is balanced by an outer field of negative particles, of much greater extent in space but of an equal value in terms of numbers of units of charge, to those found within the nucleus.

Positive and negative electricity appear as two opposing types of energy and yet are capable of representation by the unifying simple harmonic wave form or 'sine' curve; a form which in the physical universe represents the basic mode of manifestation of these various energies, so that it might well represent the behaviour of Fohat, the One Force, itself. Seen as a whole the 'sine' curve is universal and uniting but at the peaks or troughs of the waves it appears differentiated and varied. Thus Fohat may appear at a given level in a multiplicity of aspects, just as we find the energy we term electricity manifesting in various ways, so that we may use it to light or to heat a room, to emit sound or create images, or to move matter by magnetic relay action. Indeed so striking is the parallel that it is completely reasonable to find electricity termed the physical plant manifestation of Fohat, the supreme motivator. *(Reference 4 above)*

Acting as a link from one level to another, Fohat is said to be a uniting power *(References 2 and 3 above)*, since each inner level, 'seemingly subjective', acts on the level outer to oneself, 'seemingly objective', and propels it to action. Thus the manifestations of Fohat at one level, the 'operated upon' for that level, are in their turn the 'operators', acting upon an outer level and stamping their characteristics upon it. Both aspects of 'operator' and 'operated upon' are Fohat. We therefore obtain a picture of Fohat on a wide scale as a dynamic pulsating energy, not unlike electricity, which appears in the physical universe as a rhythmic pulse of positive and negative impulses, appearing to emerge from an inner level.

But Fohat is also described as vital and living and is personified as 'He'. This gives the picture of some vast and cosmic consciousness operating as this Fohatic force, playing at every level of the cosmos, and linking each form-plane to the other, spirit to mind and mind to matter. 'He', therefore, creates by his ensouling of matter, indwelling as a dynamic force. In this sense 'He' creates the laws of the universe *(Reference 4)*, for what we term laws are but our perceptions of the relationships within the cosmos. We are accustomed in our thinking to consider a law as something imposed upon matter from without, but actually it is the expression of the relationships which inhere in Nature. At the physical level, for example, because energy manifests as positive and negative electricity, the quantity and the location of its manifestation creates the laws of electrical interaction. If positive and negative electricity are seen as positive and negative phases of the manifestations of Fohat, the law of their interaction is a resultant of the activity of Fohat in the physical universe.

The particular form any manifestation of Fohat may take appears to be conditioned by the complexity (or otherwise) of the Kingdom of Nature, or level at which it is acting. Thus in the dense mineral kingdom we know it mainly as electricity in all its forms. When seen as the energiser of living material it appears in addition to its electrical manifestation as vital energy or prana playing into the physical form from a more subtle level and organising it as living material. *(Reference 5).* In the sentient animal form it is more complex still, for superimposed upon the above two manifestations is a third, that of the creative reproductive power of Kundalini. All these three, though possibly modified from inner levels, are physical plane manifestations of Fohat, the difference between them being occasioned by the complexity of the organism concerned. Or perhaps rather the complexity of the organism is the resultant of the interweaving of the three modes of manifestation of Fohat. Thus we see how, even at the physical level alone, the manifestations of Fohat are many and complex, a complexity appearing even greater when it is realized that at the mental and spiritual levels we may have to recognize the same three specialised types of Fohatic force.

Surveying the various references to Fohat in *The Secret Doctrine*, we are brought face to face with a great paradox. From one aspect one forms the impression of Fohat as an irresistible, impersonal power relentlessly driving the universe towards *its* inevitable destiny. Yet Fohat is also described as 'the Spirit of Life-giving in *His* capacity of Divine Love'. In other words, Fohat, the divine urge towards progress, does not compel but lovingly guides the evolutionary processes.

Again, Fohat is called the Pervader. He is the agency that shapes every form in creation, from worm to man, from atom to fiery sun. He, too, infuses every form with life. But beyond this, perpetually pervading all creation, he maintains it in dynamic equlibrium as a whole living entity.

In all these aspects Fohat may be identified with the One Life.

CHAPTER 2

MODES OF MANIFESTATION

by Edward L. Gardner

'Fohat is closely related to the One Life.... In its totality...
it ... is the Logos of the Platonists, and the Atman of the
Vedantins.'
(The Secret Doctrine, I. 170)

THERE ARE SOME three hundred references to Fohat in *The Secret Doctrine*, often presenting widely different descriptions of its activities, due, of course, to the innumerable ways that Fohat, the One Force, manifests through forms. In its primal undifferentiated essence Fohat is synonymous with the One Life, the Absolute. As such it must be regarded as neither conscious nor unconscious, neither benevolent nor malevolent, utterly neutral though of immeasurable power.

The One Life, Fohat, even in its highest and undifferentiated state must not be confused with any conception of deity, of 'God'. Every living and conscious being, from the loftiest to the humblest, from the very Rulers of Creation to the lowliest of creatures, is a manifestation of the One Life. The forms through which that Life functions determine its play.

AN ELECTRICAL ANALOGY

One of the many references mentioned describes Fohat as Cosmic Electricity and, since we know something of electricity and its many varied expressions, it serves well as an illustration and analogy.

Electricity from the distribution grid is reduced to lower voltages for domestic use. Differences in voltage and current depend first on the transformer that is used and then on the final form — the lamp, the stove, the motor — through which the current passes. This practice of reducing the great power of the electric current on the grid to the familiar manageable voltages suitable for our houses vividly illustrates the reduction of the enormous power of Fohat, the One Life, to the lesser values needed in constellations and solar systems and their planets.

An important point is that, just as electricity is always exactly the same in kind and quality whatever the degree of pressure may be, the One Life, Fohat, is in essence always the same, 'yesterday, today and for ever'.

The variable degrees of Fohat manifesting in systems and worlds are mentioned on page 199, Volume I:

> '...each world has its Fohat... there are as many Fohats as there are worlds, each varying in power and degree of manifestation.'

FOHAT STEPPED DOWN

The galaxy of stars, known as an Island Universe, within which our solar system exists, is the largest volume of space that we need contemplate as being within our ken. These myriads of stars and their companions are regarded, in the light of occult research, as collectively graded in three groups, each group being within a ring-pass-not with its own standard pressure of Fohat.

The focal centres of the three groups are said to be — the first and loftiest, ALKYONE of the Pleiades; second, SIRIUS; and third, HELIOS, our Sun. In other words, ALKYONE transforms Fohat from whatever the higher pressure beyond may be. SIRIUS and many others, a vast number, of equal or near status, receive from ALKYONE the Ray needed for themselves and the systems associated with them. And HELIOS, one among the many under SIRIUS, receives the still further transformed Fohatic Power for our solar manifestation.

Thus, greatly modified, this power-ray of Fohat, still of immeasurable potency, becomes the life-force received within the protective ring-pass-not of our Solar Lord, and constitutes the basic energy manifesting in Life and Form within the solar system. That Ray provides in super-abundance all the energy that can ever be used. From our human point of view and experience, no power can be created anew—power simply is.

Fohat Coiled Up

During the first three rounds on our Chain of globes, the Devic manipulators of Fohat were chiefly concerned with the mineral, plant and animal forms respectively. Also, at a quicker tempo, during the first three root-races on our earth in this the fourth round, all these forms were gradually invested with the denser physical material here available. And, as a result of the prodigious vegetation—plant growths during the second round, with tall antennae rising sky-high from a semi-fluid earth, amplified during our second racial period with lofty trees of denser growth—the earth became, and is, a charged Fohatic battery of immense power mostly contained, coiled up, in mineral form.

Throughout the involutionary arc of descent into forms, the directing hand has been almost entirely that of the Devas. Though *builders* of forms they have no stable form of their own. Their 'bodies' are illusive. (*The Secret Doctrine* III, 270) Devas adopt the 'loose far-flung forms of wind-storms, sea-waves, dust-clouds, flaming fire' and, for the infinitely varied more concrete work of form-building in the kingdoms of nature, the outer ranks of devas and nature-spirits control and shape the elemental essence of the planes by means of magnetic lines of force around their own spark-like nuclei—focal centres of life.

The aptitude and superb skill in form-building displayed by the devic hosts arises from countless cyclic recapitulations, an acquired skill that we tend to regard merely as nature's routine. In terms of the material element earth, their triumph in form-construction would seem to be this physical world, this store-house of power, this unique fourth globe of our septenary Chain.

Forms in the mineral kingdom appear to have achieved maturity and are now beginning to break down, slowly, and yield their coiled up power towards the evolution of the plant, animal and human kingdoms on the upward arc of the cycle.

It is clear that whatever power is consciously released by human labour, whether through the fiery energy of burning fuel, the momentum of falling water, or through the recently discovered process of atomic fission — all really are due to the release of Fohat from storage.

FOHAT ABOVE AND BELOW

The material growth and the health of plant and animal forms, including the human, depend on a judicious balance being maintained between terrene and solar forces — between Fohat released from below and Fohat from above. These two manifestations of Fohat, dual and divided in our present human experience, are described in *The Secret Doctrine* as Light and Life (Vol. V, 491).

> 'Above, LIGHT; Below, LIFE. The former is ever immutable, the latter manifests under the aspects of countless differentiations.'

Power that is released from below, from that which is coiled up in the physical material of the planet, is called kundalini and four layers of this are, in some measure, in functional activity through animal and human physical bodies. These constitute the wholesome fiery life from below, which is moderate and, if unobstructed and automatic, robustly healthful. The three further interior layers become *naturally* functional in the later rounds. Fohat, as kundalini, is Life from below.

Power that is from above, and which can be accepted and used by Man alone, individualised and wholly self-conscious, is the Fohatic Light, the Breath of the Most High.

The atmic permanent atom is the entrance portal for this specific Fohatic force of the above. A minute portal is this, the atmic gateway, and well symbolised by the geometric point, as appropriately as the dense physical body is symbolized by the contours of the sphere.

Our human hierarchy is said to be, in terms of consciousness, halfway between the above and the below, that is, with the awakening of mind — at mid-manas or just beyond.

The Chain of globes with which our humanity is concerned is five-fold, being built on five planes, the atmic, buddhic, mental, astral and physical — and the human constitution also has five corresponding principles, active or latent. Two of these planes and principles are dominantly of life, the buddhic, and the astral, and three are dominantly concerned with form, the atmic, mental and physical.

Fohat enters this five-fold system and is conditioned by the nature of the formal material on the planes, and is thus between two boundaries. There is a boundary, or ring-pass-not, within, at the level of the atmic gateways, and one without at the dense level. These are called ring-pass-nots because human consciousness, until full maturity is achieved, does not pass beyond them.

ABOVE AND BELOW: CORRESPONDENCES

> 'Man contains in himself every element that is found in the universe. There is nothing in the Macrocosm that is not in the Microcosm.'
>
> *The Secret Doctrine;* V, 556

> 'Man is that being in the universe . . . in whom highest spirit and lowest matter are joined together by Intelligence, thus ultimately making a manifested God . . . Every being in this universe must pass through the human kingdom . . .'
>
> *The Pedigree of Man: Annie Besant*

> '. . . Man, the direct heir of the highest Aeon . . .'
>
> *The Secret Doctrine,* V, 449.

Modes of Manifestation 15

These statements are but a few among the many in the literature of occultism that assert and stress the immense importance of the human kingdom. And the constitution of Man, with potentially 'a Mind to embrace the Universe' as a stanza in the second book of Dzyan promises, may justify the further statement that 'Man tends to become a god' (I, 214).

The means whereby Man may hope to 'embrace the universe' is in virtue of the dual nature of the mind — and one of the regular solids, the octahedron, provides us with a wonderfully lucid symbol of the human mind. Each of the regular figures of geometry has an appropriate correspondence with a human principle as follows, the point and the sphere representing the two ring-pass-nots:

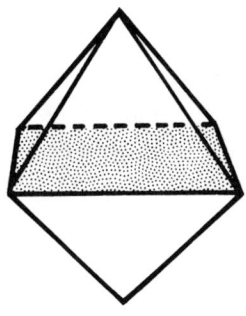

(1) Point
    ATMA I

(2) Tetrad 4 facets
    ATMA II

(3) Dedecahedron 12 facets
    BUDDHI

Sphere (7)
PHYSICAL II

Cube 6 facets (6)
PHYSICAL I

Icosahedron 20 facets (5)
ASTRAL

(4)
Octahedron 8 facets
MANAS

I — Higher;   *II — Lower*

Diagram I

The eight-sided figure, the octahedron, in the 4th or middle position above, which corresponds with the dual mind, is composed of two pyramids having a common base, one above and one below the central square. The common base, shared by both the upper and inverted pyramids, symbolises middle manas. This is the diaphragm or screen, between the higher and the lower mind, shared by both. Thus the screen has, so to speak, an upper face and an under face. The screen at middle manas serves as a mirror, for it can reflect — to man's probing consciousness — impressions sent in by the sense-organs from the physical world, as also from the spiritual realm above. The upper face is the mirror that can record an intuition.

On the involuntary journey from point to sphere, from atma to the physical world — the path of forthgoing — the interest of humanity is fostered and encouraged, by the devic hosts of Lucifer, by the contacts made as mirrored on the under face of the screen. The Asura Hierarchy, highly skilled at the lower mental level, assists on this, the downward arc, that humanity be persuaded, enticed, tempted to push out to the 'sphere' of the physical world. Further to ensure this, the 'above' is veiled.

> 'He shuts out the Above and leaves the Below . . . '
> I *Stanzas of Dzyan*, 3, vii.

Shutting out the Above led to that which is described in occultism as the deep sleep of the spiritual triad, the permanent atoms of atma-buddhi-manas. Through the latter part of the third root-race and throughout the fourth root-race, the momentum towards interest in the external world of the Below, imparted by the Luciferian devas and the Asura hierarchy, continued. There is still plenty of evidence of its strength. The bridge linking the Below and the Above is now a-building.

The three major aspects of Fohat in relation to our humanity may be summarized as follows:

(1) *From Below.* Kundalini, the coiled-up force within

the physical material of the mineral kingdom of form. Four layers released at present and a fifth partially. *(Third Aspect)* Focal centre of distribution, the muladhara chakra.

(2) *Of the Open Spaces.* Prana of all planes. Vitality globules of the atmosphere assembled by the devas in the sun's rays. *(Second Aspect)* Focal centre for distribution, the spleen chakra. It is this Prana, specialized, that is conveyed in the practice of vital magnetic healing by the competent practitioner.

(3) *From Above.* Fohat, the 'Divine Breath', specialized for the purpose of the Scheme. 'Coloured' is the term often used to describe the modifiation of the thread of Fohat that plays in humanity through the atmic permanent atoms — a fiery thread of immense potency. *(First Aspect)* Focal Centre: The Crown.

> '... said the Flame to the Spark "Thou art myself, my image and my shadow. I have clothed myself in thee..."'
> I *Dzyan* VII, 7

CHAPTER 3

FOHAT AND INTUITIONAL RESEARCH
by Edward L. Gardner

THE FORM SIDE

THE THREAD OF FOHAT from above is buddhic light, the 'light that lighteth every man that cometh into the world' and this is focused by the higher mind. In man the light becomes a concentrated point of brilliance that plays on the screen of the lower mind much as light that enters the physical eye is focused by the lens and then plays on the retina. The eye is indeed a physical correspondence of the dual mind.

The ability to focus the buddhic light accompanies individualization, being due to the awakening of the mental permanent atom in place of the mental unit of the fourth sub-plane. The change marks a distinction between the younger kingdoms and the human, from a mere awareness or consciousness of living—to self-consciousness. The difference may be illustrated by that between rays of light passing through a flat pane of glass and similar rays passing through a convex lens. Through the former, the rays are unconcentrated, unpointed, diffused; through the lens the rays are condensed, collected, acute. Because buddhi-manas thus functions as a point of brilliant light, it is regarded as of the formless world: the lower mind is an assembled body, a three-dimensional screen of exquisite receptivity, a subtle mental retina. The interplay and friction between the focused light and the mental screen date the birth of self-consciousness.

The attention of human consciousness is for many incarnations almost wholly directed to the pictures reaching the mind from the outer world by means of the sense-organs — due, as above described, to the stimulation of humanity's Divine Instructors. These all impinge upon the under-face of the mental screen, nor have we yet exhausted the possibilities of this under mirror, as an analysis will show.

A physical object seen by the eyes is conveyed to and reproduced in the mental body as a three-dimensional form. The surfaces only of such a form are seen, although the interior may also be reproduced in the mind if the etheric 'eyes' function. These are delicate membranes that surround and almost enclose the denser organ, something like a shade behind an electric light. They do not as yet usually function, perhaps fortunately, because unless under complete control they merely blur and confuse the vision. When they do function, then etheric clairvoyance is experienced and the interior of the reproduced object is also seen, for the flashing point of focused light then explores the whole. So incredibly swift is its movement that depth is absorbed and the interior of the solid object appears as a succession of flat planes. It is this latent ability that probably, even now, endows sight with perspective — for the simple retinal image of the eye is itself flat.

A propos, x-ray photography in medical diagnosis provides print copies of surfaces at varying depth in the physical body, and can reveal flaws in large metal castings.[*] Another example, this, of an externalised faculty — etheric sight — which at present is latent in humanity.

'The microcosm corresponds in every particular to the macrocosm.'

The high speed in action of the point of consciousness in practical life may be appreciated if we notice how easy it

---

[*]More recent techniques use infra-red light, ultrasonics or nuclear magnetic resonance in scanners that enable us to "see" deeply into material bodies.

is for quite an average mind to re-view a number of memories, separate memory records, link them together and reach a judgment and decision. For example, one sees an object, say a rose in full bloom, and an opinion is asked as to its quality and name. Many memory pictures of roses, in the mind, are re-viewed, a comparison is made with the new picture, and a judgment and the rose's name may, almost instantly, follow. The concrete mind, although actually a three-dimensional body, is well symbolised by a two-dimensional screen because of the speed of the play of consciousness. Hence the aptness of the flat square common to the two pyramids of our octahedral figure — with an upper and an under face.

In this example, and all similar cases, the under-face mirror is dominant. In the exercise of the intuitive faculty, the upper face is used but, naturally at present, not so readily. But long before etheric and astral sight are efficiently developed, the mind can be trained to record an impression that originates from the Above — and such an impression of truth is well called an intuition. We can examine the mental mechanism more closely.

Atma, from which streams the Fohatic thread of buddhic light, is a static principle. That is, it is relatively static in relation to the personality, in the same sense that the sun is relatively static to the planets, or that a broadcasting-station is static to the many radio-sets tuned in to it.

The need in any experiment involving the intuition is that the personality should be quietly expectant, receptive and alert — and hence 'in line' with a static centre. Sometimes, however, a clear intuition may be registered even when under strain, as a storm-tossed ship may momentarily catch a beam from a lighthouse. The reason for such a single glimpse may be that the emotional tension of an exacting situation has keyed the personality into, temporarily, a unit — and the upper mirror is momentarily clear.

The technique required in order to record, at will, an

impression from buddhi-manas on the mid-mental screen, seems to involve a change in the lens. This change shortens the higher mental focus which, usually, plays through to the brain area as the background for lower mental activity. Such a shortening means that the point of consciousness plays upon the upper face of the mental mirror and not beyond.

Through long practice, consciousness probes and explores the field of the lower mind easily and rapidly and covers a wide area varying much in clarity and diffusion. To register impressions from above, the focus is, so to speak, just above the screen. The light is more intense and the field is more localised.

The mental permanent atom, the true lens of the buddhic light, has also an aura of the second and third sub-planes of conditioned manas. This fringe tends to soften and slightly diffuse the impression received — at present, fortunately so!

The shortening of focus means that the attention is drawn above the screen, if only for an instant, and the range of its perceptive clarity then is that of the archetypal world and the brilliantly vivid first elemental kingdom. Here are the archetypes and sub-archetypes, the blue-prints of reality, and, given this technique, they are within human reach.

It is taken for granted that the student will be aware of the disciplines relating to the personal nature that are recommended to ensure that conditions are suitable for the practice of this technique. The essentials are considerable self-control, pure intentions and quietude.

INTUITIVE RESEARCH: THE LIFE SIDE

It may be useful to describe in detail an experimental effort along the lines indicated.

Let the student select a study that interests him and read it up and ponder over it from as many angles as possible. The subject chosen may be a poetical passage, a verse from a scripture, a symbolic figure, a mathematical problem, an obscure theory of science, no matter what — provided that, in the early stages, the choice is of real interest.

Fill the mind with that which relates to the chosen subject for a short time on each of several days — unless of course it is already thoroughly familiar. Such a study will attract the mental archetype of which the chosen subject is a limited extension — because nothing can be thought of relative to ideals that is not already extant on the higher plane.

In parenthesis, a definition of an ARCHETYPE is worth noting. The dictionary states that 'an archetype is the original model or ideal after which anything is made'. To illustrate with a mundane example — you may see in some magazine a quotation from some poem of a few lines, the whole of which you would like to know. A letter to the editor leads to the poem being printed in full in the next issue. The interest shown in the few lines and the letter of enquiry led directly to the poem, the archetype, coming your way. The previous study of a subject, mentioned above, corresponds to the lines of the poem that were first seen. Translate this into the terms of the higher mental plane, for the correspondence is close; the thoughts of clear thinkers attract their like.

After the preparation relating to the subject chosen, appoint a time with yourself, say during the following day, preferably soon after awakening, for the experiment. (Your own mind-elemental will, by this time, be interested.)

When the time arrives, be alone, undisturbed and quiet. Call to mind the subject of the special study, merely that, nothing more. This is in the nature of a very brief review and for a few moments only. Then, positively withdraw. By withdraw I mean make a wilfull attempt to rise from the lower mental field — yet remain receptive. Think of nothing, no thing, no think.

Some resistance may be expected to this effort but, quite gently, insist that your will must be obeyed.

The word 'wilful' must be taken as meaning exactly

what it implies — an act of will. Will is not of course power itself — but it is the direction of power, corresponding to wilfully moving a switch that controls a current of electricity.

If there be a query as to how this is done, ask yourself how you change the shape of the lens of the eye when you wish to see out of the window as you take your eyes from a book. The lens moves and is altered in shape in order that the focus for the outside view be rectified. The explanation of the seemingly unconscious effort is that you merely issue a tacit order — the trained devic life, in charge of the elemental life of the eye, obeys.

When successful in the experiment above detailed, and the wilful order is given, then the associated counterpart of the subject of study, the archetype or sub-archetype, will, in perhaps the briefest of flashes, illumine the mind's content.

A successful result appears to be something like a depth projection — a fulness of understanding arrives and a perspective view of the subject studied will be available to consciousness. This may easily take an hour or much more to absorb — and then extend in detail.

CHAPTER 4

FOHAT

By Josephine Ransom

THE FIRST MENTION OF FOHAT is in Stanza of Dzyan 3, shloka 12: 'Then Svabhavat sends Fohat to harden the Atoms.' In what follows some repetition is necessary and unavoidable to get a clear notion of what Fohat is. The word Fohat is a Buddhist term and denotes many things, many displays of the One Life presenting itself in countless ways.

Trying to trace the nature and function of Fohat takes us from pure abstractions through the regions of the formation of the innumerable great or small universes which are aggregations of forces, to the atomic states of our universe.

First, let us consider some of the abstractions. Svabhavat is that Father-Mother (so difficult to comprehend) who 'spin a web,'[1]* and 'this Web is the Universe spun out of the Two Substances made in One, which is Svabhavat.'[2] Born of the relation of the two is Fohat, which emerges out of the 'Darkness' to set Spirit and Matter into conscious relationship — thus becoming Their Son, the cold flame born of the Light of Matter, Mother, and the Fire of Spirit, Father. In its own nature, Fohat is beginningless and endless.

Fohat dissociates and scatters the Atoms, the Sons, 'which expand and contract through their own hearts.'[3] For

---

*The list of references will be found at the end of the chapter.

these Atoms are those which have been called 'holes dug in Space.' They expand when the 'Breath of Fire is upon them,'[4] which means that no limit is set to their interior movement, but the 'Breath of the Mother'[5] sets limits or contracts them by the terrific momentum of the solidity or mass of eternal Matter-Mother, Substance, which resists the expansion. This pressure 'hardens the atoms'[6] by limiting them to definite size and shape. And 'each is part of the Web . . . and each becomes in turn a World,'[7] for each has in it the nature of the 'Self-Existent Lord' Whose Breath Fohat is. And here, too, is that *Man* Pattern, the 'Force or Divine Man, the Sum Total.'[8]

After certain cosmic states are established then comes the stage when the 'Primordial Seven'[9] (the Dhyan Chohans) are stationed in their Great Cosmic Duties and they embody in themselves the Atoms to which They, in turn, give a special movement and the force of Fohat is now turned in its circumscribed area and becomes the 'fiery Whirlwind,'[10] running 'circular errands'[11] within itself. (See atomic structure, given first in the *Ancient Wisdom*, by Dr. Besant, and then in other books.)

'Fohat is the Steed and Thought is the Rider,'[12] or, Fohat is the One Life, for he brings about the vast changes that we call unmanifestation and manifestation. He moves from being the 'Unknown One, the Infinite TOTALITY,' into being 'the Manifested One, or the periodical Manvantaric Deity.'[13] He *is* the Universal Mind which is the Creative Logos, four-natured because triple and yet synthesized as One. He calls into activity the 'Hosts of the higher creative Dhyan Chohans'[14] and He, the Voice within each Spark or Atom, joins them together in vast aggregations and calls out of the deeps the Lipika to circumscribe karmically the nature of the field of the new manifestation. Also the power of Fohat comes into action as the One Supreme and Eternal Wisdom, and is by 'the action of this Manifested Wisdom, or Mahat — represented by these innumerable centres of spiritual energy in the Kosmos — the Reflection of the Universal Mind. . . .'[15]

Fohat calls to 'the innumerable Sparks and joins them together,'[16] for he is at once the *manifested* substance, the One Element, and also composes all the differentiations which in their combinations make the seven principles of Akasha, and—by differentiating it into various centres of energy—sets in motion the law of Cosmic Evolution, the seven Planes with all their activities.

'Fohat, then, is the ... transcendental binding unity of all cosmic energies, on the unseen as on the manifested planes, the action of which resembles—on an immense scale—that of a living Force created by WILL.'[17] In this mode of activity Fohat performs the function of 'Pervader'. He is Vishnu—Vish, the Pervader, 'the Manufacturer, because he shapes the atoms from crude material'[18] (i.e. the 'holes' he has already in his creative aspect dug in Space). On the Cosmic level Fohat is present in the constructive power that carries out the formation of things, following the plan in the mind of Nature, or in the Divine Thought (his own). He is the messenger of a cosmic and human Ideation, the active force in universal life.

It is most significant that Fohat is described as striding through the seven regions of the Universe in three steps; thus revealing the esoteric doctrine that the One Fohat in His Three aspects is the triple controlling Power of the Logos as Will, Wisdom and Activity—Shiva, Vishnu, Brahma—Father-Mother, Son and Holy Ghost. In theosophical literature we find the operation of these three aspects explained with great lucidity, stating plainly what with difficulty one finds in *The Secret Doctrine*, though one must say that in *The Secret Doctrine* there is a richness of suggestion in each statement which all other literature seems to lack.

Then comes another stage in the objectivization of Fohat, the objectivized Thought of the Gods, the 'Word made Flesh,' for the Three and Seven 'Strides'[19] refer to the seven cosmic spheres inhabited by man as well as the seven regions of the earth. The three strides are the activities of the three Logoi and the seven planes of the great Kosmos, and their repetition or reflection in the seven planes of our small Cosmos. 'The three strides relate metaphysically to the descent of Spirit into Matter, of the Logos falling as a ray into the Spirit, then into the Soul, and finally into the human form of man, in which it becomes Life.'

This passage is important for it plainly indicates that Fohat, the One Life, is the Spirit which becomes, within Substance, the Matter or Energy of a Universe, its Logos. Then this One-yet-threefold Logos falls as a triple ray into the Spirit, or rather, forms another phase of activity of the One on the way to manifestation. Then, the First Logos of this triple Ray becomes one of the aspects of the World Soul, a process that is repeated when the First Logos unites with the other two aspects of Himself at the formation of a human Soul. The three thus unite in evolution and make again a triple Oneness; and this is the stage peculiar to MAN.

It is at this stage that the 'Life,' or Fohat, becomes the hidden Kundalini within the framework of his physical nature. The three are conjoined as Sushumna, Ida and Pingala and are the everlasting 'call' of the One to Self-Realization. It is to be noted here that it was to this work of Self-realization that H.P.B. called Fifth Root Race mankind.

In the cosmic formation of MAN, which human man repeats, the Principle known as the Divine Soul or *Buddhi*, though a mere breath, in our conception is still something material when compared with Divine Spirit (Atma) of which it is the carrier or vehicle. 'Fohat in his capacity of DIVINE LOVE . . . is shown . . . trying to bring the pure Spirit, the Ray inseparable in the ONE Absolute, into union with the Soul, the two constituting in Man the MONAD, and in Nature the first link between the ever-unconditioned and the manifested.'[20]

It is not easy to realize that *all* the stages of manifestation are becomings, not changes, that the One always IS, but can and does exhibit Itself in many ways—we might say a certain number of ways, the laws of a given universe—and is never other than itself whether described as Spirit and Matter, or as First, Second or Third Logos, or as the Dhyan

Chohans, or the Lipika, or the Gods in their Imperial Robes who rule the Elements, the 'Garment of God', or as Man, the truly powerful reflection of all that *Is*. He, Fohat, is the Eternal and the ephemeral, the Immortal Changeless and Timeless, and is also brief mortality and the split second of time.

In Cosmic activities Fohat is described as the 'Light of the Logos,'[21] the hidden 'spirit' of ELECTRICITY, which is 'the LIFE of the Universe.'[22] The Logos, triple in nature, shows himself as objective matter, subjective thought and the link between them. 'The whole manifested solar system exists in its Sukshma (subtle) form in this light or energy of the Logos...' while the whole cosmos in its objective form is the Word of the Logos. In fact Fohat exists in all the conditions of the solar system, whether as the One, the Three, the Four or the Five or the Seven of all the changes that are rung by the One. 'On the Cosmic plane he is behind all such manifestations as light, heat, sound, adhesion, etc.[23] 'As an objective and evident Reality, we speak of a septenary scale of manifestation, which begins at the upper rung with the One Unknowable CAUSALITY, and ends as Omnipresent Mind and Life, omnipresent in every atom of Matter.'[24]

From this point we can study the movements of Fohat in all the energies of Nature, and by studying the phenomenal universe attempt to understand the nature of the noumenal. For the energies of Nature are defined by Occult Science as 'supersensuous effects in their hidden behaviour and as objective phenomena in the world of sense.... They all pertain to, and are emanations of, still more supersensuous spiritual qualities, not personated by, but belonging to, real and conscious CAUSES.'[25]

There are many other most interesting statements in *The Secret Doctrine* about the operations of Fohat in its manifold permutations. The few we have considered are sufficient to help us to perceive the simplicity of the nature of the One Life, which yet contains within itself all the multifarious operations in our own and other Universes. It is indeed all important to know the One in the Many. How to do this is the basic instruction in all Occult Teaching.

List of References — Chapter IV:
[1] *Stanzas of Dzyan* III, shloka 10.
[2] Ibid.
[3] *Stanza* III, 11.
[4] Ibid.
[5] Ibid.
[6] *Stanza* III, 12.
[7] Ibid.
[8] *Stanza* IV, 3.
[9] *Stanza* IV, 2 and V, 1.
[10] *Stanza* V, 1.
[11] *Stanza* V, 2.
[12] Ibid.
[13] *The Secret Doctrine*, I, 170.
[14] Ibid.
[15] Ibid.
[16] *Stanza* V, 2.
[17] *The Secret Doctrine*, I, 170.
[18] *The Secret Doctrine*, I, 171.
[19] *Stanza* V, 2.
[20] *The Secret Doctrine*, I, 178.
[21] *The Secret Doctrine*, III, 399, Footnote 4.
[22] *The Secret Doctrine*, I, 171, 195.
[23] *The Secret Doctrine*, I, 195.
[24] *The Secret Doctrine*, I, 196.
[25] *The Secret Doctrine*, I, 201.

CHAPTER 5

REFERENCES TO FOHAT IN THE SECRET DOCTRINE

Arranged consecutively by E. LESTER SMITH

The following extracts relating to FOHAT, taken from *The Secret Doctrine*, are shorn, as far as possible, of confusing parentheses and mythological terms and are rearranged to form a reasonably consecutive account of the nature of FOHAT.

References are given both to the Adyar and the Third Edition, (A) indicating the Adyar Edition, (B) the Third.

Just as pre-cosmic Ideation is the root of all individual Consciousness, so Pre-Cosmic Substance is the substratum of Matter in the various grades of its differentiation. (A) I, 81; (B) I, 43.

The Manifested Universe, therefore, is pervaded by duality, which is, as it were, the very essence of its *Ex*-istence as 'Manifestation'. But just as the opposite poles of Subject and Object, Spirit and Matter, are but aspects of the One Unity in which they are synthesized, so, in the Manifested Universe, there is 'that' which links Spirit to Matter, Subject to Object.

This something, at present unknown to Western speculation, is called by Occultists Fohat. It is the 'bridge' by which the Ideas existing in the Divine Thought are impressed on Cosmic Substance as the 'Laws of Nature'. Fohat is thus the dynamic energy of Cosmic ideation; or, regarded from the other side, it is the intelligent medium, the guiding power of all manifestation, the 'Thought Divine' transmitted and made manifest through the Dhyan Chohans, the Architects of the visible World. Thus from Spirit, or Cosmic Ideation, comes our Consciousness, from Cosmic Substance the several Vehicles in which that Consciousness is individualised and attains to Self—or reflective—consciousness; while Fohat, in its various manifestations, is the mysterious link between Mind and Matter, the animating principle electrifying every atom into life. (A) I, 81; (B) I, 44.

Fohat, being one of the most, if not the most important character in esoteric cosmogony, should be minutely described. As in the oldest Grecian cosmogony . . . so Fohat is one thing in the yet Unmanifested Universe, and another in the phenomenal and Cosmic World. In the latter, he is that occult, electric, vital power, which, under the Will of the Creative Logos, unites and brings together all forms, giving them the first impulse, which in time becomes law. But in the Unmanifested Universe, Fohat is no more this, than Eros is the later brilliant winged Cupid, or Love. Fohat has naught to do with Cosmos yet, since Cosmos is not born, and the Gods still sleep in the bosom of 'Father-Mother.' He is an abstract philosophical idea. He produces nothing yet by himself; he is simply that potential creative Power, in virtue of whose action the Noumenon of all future phenomena divides, so to speak, but to reunite in a mystic supersensuous act, and emit the creative Ray. When the 'Divine Son' breaks forth, then Fohat becomes the propelling force, the active Power which causes the One to become Two and Three—on the cosmic plane of manifestation. The triple One differentiates into the Many, and then Fohat is transformed into that force which brings together the elemental atoms, and makes them aggregate and combine. (A) I, 169; (B) I, 134.

Fohat is closely related to the 'One Life.' From the Unknown One, the Infinite Totality, the Manifested One, or the periodical Manvantaric Deity, emanates; and this is the Universal Mind, which, separated from its Fountain-Source, is the Demiurge or the Creative Logos of the Western Kabalists . . . . By the action of the Manifested Wisdom, or Mahat, the Reflection of the Universal Mind, which is Cosmic Ideation, and the Intellectual Force accompanying such Ideation becomes objectively the Fohat of the Buddhist esoteric philosopher. Fohat, running along the seven principles of Akasha, acts upon manifested Substance, or the One Element, as declared above, and, by differentiating it into various centres of energy, sets in motion the law of Cosmic Evolution, which, in obedience to the Ideation of the Universal Mind, brings into existence all the various states of being in the manifested Solar System. (A) I, 170; (B) I, 135.

The Primordial Seven, The First Seven Breaths of the Dragon of Wisom, produce in their turn from their Holy Circumgyrating breaths the Fiery Whirlwind. Stanzas of Dzyan, V, I.

The 'Fiery Whirlwind' is the incandescent cosmic dust which only follows magnetically, as the iron filings follow the magnet, the directing thought of the 'Creative Forces.' Yet, this cosmic dust is something more; for every atom in the Universe has the potentiality of self-consciousness in it, and is, like the Monads of Leibnitz, a Universe in itself, and *for* itself. *It is an atom and an angel.* (A) I, 167; (B) I, 132.

Then Svabhavat sends Fohat to harden the atoms. Each is a part of the web. Reflecting the 'Self-Existent Lord' like a mirror, each becomes in turn a world. Stanzas III, 12.

Primordial Matter, then, before it emerges from the plane of the never-manifesting, and awakens to the thrill of action under the impulse of Fohat, is but 'a cool radiance, colourless, formless, tasteless, and devoid of every quality and aspect.' (A) I, 147; (B) I, 110.

Fohat hardens the atoms; i.e., by infusing energy into them, he scatters the 'Atoms,' or Primordial Matter. *'He scatters himself while scattering Matter into Atoms.'*

It is through Fohat that the ideas of the Universal Mind are impressed upon Matter. (A) I, 150; (B) I, 113.

They make of him the messenger of their will. The Dzyu becomes Fohat: the swift son of the divine sons, whose sons are the Lipika, runs circular errands. Fohat is the steed, and the thought is the rider . . . . Stanzas V, 2.

This shows the 'Primordial Seven' using for their Vehicle . . . Fohat, called in consequence, the 'Messenger of their Will' — the 'Fiery Whirlwind'.

'Dzyu becomes Fohat' — the expression itself shows it. Dzyu is the one Real (Magical) Knowledge, or Occult Wisdom; which, dealing with eternal truths and primal causes, becomes almost omnipotence when applied in the right direction. Its antithesis is Dzyu-mi, that which deals with illusions and false appearances only, as in our exoteric modern sciences. In this case, Dzyu is the expression of the collective Wisdom of the Dhyani-Buddhas. (A) I, 168; (B) I, 133.

Fohat runs the Manus' (or Dhyan Chohans') errands, and causes the ideal prototypes to expand from within without — that is, to cross gradually, on a descending scale, all the planes, from the noumenal to the lowest phenomenal, to bloom finally on the last into full objectivity — the acme of Illusion, or the grossest matter. (A) I, 132; (B) I, 93.

Some faint idea of the nature of Fohat may be gathered from the appellation 'Cosmic Electricity', sometimes applied to it; but in this case, to the commonly known properties of electricity, must be added others, including intelligence. (A) I, 150; (B) I, 113.

Fohat, then, is the personified electric vital power, the transcendental binding unity of all cosmic energies, on the unseen as on the manifested planes, the action of which resembles — on an immense scale — that of a living Force created by Will, in those phenomena where the seemingly subjective acts on the seemingly objective, and propels it to action. Fohat is not only the living Symbol and Container of that Force, but is looked upon by the Occultists as an Entity; the forces it acts upon being cosmic, human, and terrestrial, and exercising their influence on all these planes respectively. On the earlthly plane, its influence is felt in the magnetic and active force generated by the strong desire of the magnetiser. On the Cosmic, it is present in the constructive power that, in the formation of things — from the planetary system down to the glow-worm and simple daisy — carries out the plan in the mind of Nature, or in the Divine Thought, with regard to the development and growth of a particular thing. It is, metaphysically, the objectivised Thought of the Gods, the 'Word made flesh', on a lower scale, and the messenger of cosmic and human ideation; the active force in Universal Life. In its secondary aspect, Fohat is the Solar Energy, the electric vital fluid, and the preserving Fourth Principle, the Animal Soul of Nature, so to say, or — Electricity (A) I, 170; (B), I, 136.

The ancients represented it by a serpent, for *'Fohat hisses as he glides hither and thither'*, in zigzags. . . . It is the magical agent *par excellence*, and designates in Hermetic philosophy 'Life infused into Primordial Matter', the essence that composes all things, and the spirit that determines their form. (A) I, 143; (B) I, 105.

It was Fohat to which Plato referred as that Something that 'caused the Universe to move with circular motion'. (A) I, 250; (B) I, 222.

Fohat turns with his two hands in contrary directions the 'seed' and the 'curds', or Cosmic Matter. (A) II, 397; (B) I, 736.

Gross ponderable matter is the body, the shell, of Matter or Substance, the female passive principle; and this Fohatic Force is the second principle, Prana — the male and the active. (A) II, 249, Footnote 2; (B) I, 572, Footnote.

Cupid or Love in his primitive sense is Eros, the Divine Will, or Desire of manifesting itself through visible creation. Thence Fohat, the prototype of Eros, becomes on Earth the Great Power 'Life-Electricity', or the Spirit of 'Life-giving'. (A) III, 76; (B) II, 69.

It is the action of Fohat upon a compound, or even upon a simple, body that produces life. When a body dies, it passes into the same polarity as its male energy, and repels therefore the active agent, which, losing hold of the whole, fastens on the parts or molecules, this action being called chemical. Vishnu, the Preserver, transforms himself into Rudra-Shiva, the Destroyer — a correlation seemingly unknown to Science. (A) II, 250, Footnote 3; (B) I, 573, Footnote.

The cause underlying physiological variation in

species — one to which all other laws are subordinate and secondary — is a sub-conscious intelligence pervading matter, ultimately traceable to a *reflection* of the Divine and Dhyan-Chohanic wisdom. A not altogether dissimilar conclusion has been arrived at by so well known a thinker as Ed. von Hartmann, who, despairing of the efficacy of *unaided* Natural Selection, regards Evolution as being intelligently guided by the Unconscious — the Cosmic Logos of Occultism. But the latter acts only mediately through Fohat, or Dhyan-Chohanic energy, and not quite in the direct manner which the great pessimist describes. (A) II, 219; (B) II, 685.

. . . He passes like lightning through the fiery clouds; takes three, and five, and seven strides through the seven regions above, and the seven below. He lifts his voice, and calls the innumerable sparks, and joins them together. Stanza V, 2.

He is their guiding spirit and leader. When he commences work, he separates the Sparks of the lower kingdom, that float and thrill with joy in their radiant dwellings, and forms therewith the germs of wheels. He places them in the six directions of space, and one in the middle — the central wheel.

Footnotes:

Fiery Clouds — Cosmic mists.

The Seven below — The World to be.

Sparks — Atoms.

Sparks of the Lower Kingdom — The mineral atoms.

Radiant dwellings — Gaseous clouds. *Stanza V, 3.*

The Three and Seven 'Strides' refer to the seven spheres inhabited by man, in the Esoteric Doctrine, as well as to the seven regions of the Earth. (A) I, 171; (B) I, 137.

'Wheels' are the centres of force, around which primordial matter expands, and, passing through all the six stages of consolidation, becomes spheroidal and ends by being transformed into globes or spheres. It is one of the fundamental dogmas of Esoteric cosmogony, that during the Kalpas (or Aeons) of Life, Motion, which, during the periods of Rest, *'pulsates and thrills through every slumbering atom'* — assumes an ever growing tendency, from the first awakening of Kosmos to a new 'Day', to circular movement. 'The Deity becomes a Whirlwind'. (A) I, 176; (B) I, 141.

By the 'Six Directions of Space' is here meant the 'Double Triangle', the junction and blending together of pure Spirit and Matter, of the Arupa and the Rupa, of which the Triangles are a symbol. (A) I, 177; (B) I, 143.

Fohat traces spiral lines to unite the sixth to the seventh — the crown. An army of the Sons of Light stands at each angle; the Lipika, in the middle wheel. They say: 'This is good'. The first divine world is ready; the first, the second. Then the 'Divine Arupa' reflects itself in Chhaya Loka the first garment of Anupadaka.

Footnotes:

They Say — The Lipika.

The First, the Second — That is: the First is now the Second World.

The 'Divine Arupa' — The Formless Universe of Thought.

Chhaya Loka — The Shadowy World of Primal Form, or the Intellectual. *Stanza V, 4.*

This tracing of 'spiral lines' refers to the evolution of Man's as well as Nature's Principles; an evolution which takes place gradually, as does everything else in Nature. The Sixth Principle in Man (Buddhi, the Divine Soul), though a mere breath, in our conceptions, is still something material when compared with Divine Spirit (Atma), of which it is the carrier or vehicle. Fohat, in his capacity of Divine Love (Eros), the electric power of affinity and sympathy, is shown, allegorically, trying to bring the pure Spirit, the Ray inseparable from the One Absolute, into union with the Soul, the two constituting in Man the Monad, and in Nature the first link between the ever-unconditioned and the manifested. (A) I, 178; (B) I, 144.

The radical unity of the ultimate essence of each constituent part of compounds in Nature — from star to mineral atom, from the highest Dhyan Chohan to the smallest infusorium, in the fullest acceptation of the term, and whether applied to the spiritual, intellectual, or physical worlds—this unity is the one fundamental law in Occult Science. (A) I, 179; (B) I, 145.

Fohat takes five strides, and builds a winged wheel at each corner of the square for the Four Holy Ones . . . and their armies.

Footnotes:

Five Strides—having already taken the first three.

Their Armies—Hosts. *Stanza V, 5.*

The 'Strides' refer to both the cosmic and the human Principles—the latter of which consists, in the exoteric division, of three (Spirit, Soul and Body), and, in the esoteric calculation, of seven Principles—three Rays of the essence and four Aspects. . . . The four Aspects are the body, its life or vitality, and the 'double' of the body—the triad which disappears with the death of the person—and the Kama Rupa which disintegrates in Kama Loka.

From a cosmic point of view, Fohat taking 'Five Strides' refers here to the five upper planes of Consciousness and Being, the sixth and the seventh (counting downwards) being the astral and the terrestrial, or the two lower planes.

Four Winged Wheels at each corner . . . for the Four Holy Ones and their Armies (Hosts). These are the 'Four Maharajahs' or great Kings, of the Dhyan Chohans, the Devas, who preside each over one of the four cardinal points. They are the Regents, or Angels, who rule over the Cosmical Forces of North, South, East and West, Forces having each a distinct Occult property. These Beings are also connected with Karma, as the latter needs physical and material agents to carry out its decrees. (A) I, 180; (B) I, 147.

The 'Four' are the protectors of mankind and also the agents of Karma on Earth, whereas the Lipika are concerned with Humanity's hereafter. (A) I, 185; (B) I, 151.

There are three chief Groups of Builders, and as many of the planetary Spirits and the Lipika, each Group being again divided into seven sub-groups. (A) I, 186; (B) I, 152.

The Lipika are the Spirits of the Universe, whereas the Builders are only our own planetary dieties. (A) I, 186; (B) I, 153.

By the power of the Mother of Mercy and Knowledge, Kwan-Yin — the triple of Kwan-Shai-Yin, residing in Kwan-Yin-Tien — Fohat, the breath of their progeny, the Son of the Sons having called forth, from the lower abyss, the illusive form of Sien-Tchan and the Seven Elements.
Footnotes:
The lower Abyss — Chaos.
Sien-Tchan — Our Universe. *Stanza VI,* 1.
This Stanza is translated from the Chinese text, and the names given as the equivalents of the original terms are preserved. (A) I, 193; (B) I, 160.

The swift and the radiant One produces the seven Laya Centres, against which none will prevail to the great day 'be with us'; and seats the Universe on these eternal foundations, surrounding the Sien-Tchan with the elementary germs. Stanza VI, 2.

The seven Laya Centres are the seven zero-points, using the term zero in the same sense that Chemists do. It indicates, in Esotericism, a point at which the reckoning of differentiation begins. From these Centres . . . begins the differentiation of the Elements which enter into the constitution of our Solar System. It has often been asked what is the exact definition of Fohat and his powers and functions, for he seems to exercise those of a Personal God as understood in the popular religions. . . . 'The whole cosmos must necessarily exist in the one source of energy from which this light (Fohat) emanates' . . . 'just as a human being is composed of seven principles, differentiated matter in the solar system exists in seven different conditions'. So does Fohat. Fohat has several meanings. He is called the 'Builder of the Builders', the Force that he personifies having formed our Septenary Chain. He is One and Seven, and on the cosmic plane is behind all such manifestations as light, heat, sound, adhesion, etc., etc., and is the 'spirit' of electricity, which is the Life of the Universe. As an abstraction, we will call it the One Life; as an objective and evident Reality, we speak of a septenary scale of manifestation, which begins at the upper rung with the One Unknowable Causality, and ends as Omnipresent Mind and Life, immanent in every atom of Matter. Thus, while Science speaks of its evolution through brute matter, blind force, and senseless motion, the Occultists point to *Intelligent* Law and *Sentient* Life, and add that Fohat is the guiding Spirit of all this. Yet he is no personal god at all, but the emanation of those other Powers behind him, whom the Christians call the 'Messengers' of their God . . . and we the Messenger of the primordial Sons of Life and Light. (A) I, 195; (B) I, 162.

. . . all the Forces, such as Light, Heat, Electricity, etc., are called the 'Gods' — Esoterically.

This, indeed, must be so, since the Esoteric Teachings in Egypt and India were identical. And, therefore, the personification of Fohat, synthesizing all the manifesting Forces in Nature is a legitimate result. Moreover, as will be shown later, the real and Occult Forces in Nature only now begin to be known. . . . (A) II, 397; (B) I, 735.

Fohat is the key in Occultism which opens and unriddles the multiform symbols and allegories in the so-called mythology of every nation; demonstrating the wonderful philosophy and the deep insight into the mysteries of Nature, contained in the Egyptian and Chaldean as well as in the Aryan religions. Fohat, shown in his true character, proves how deeply versed were all these prehistoric nations in every Science of Nature, now called the physical and chemical branches of Natural Philosophy. In India, Fohat is the scientific aspect of both Vishnu and Indra, the latter older and more important in the *Rig Veda* than his sectarian successor; while in Egypt, Fohat was known as Toom issued of Noot, or Osiris in his character of a primordial God, creator of heaven and of beings. For Toom is spoken of as the Protean God who *generates other Gods* and gives himself the form he likes; the 'Master of Life, giving their vigour to the Gods'. He is the *overseer* of the Gods, and he 'who creates spirits and gives them shape and life'; he is 'the North Wind and the Spirit of the West'; and finally the 'Setting Sun of Life', or the vital electric force that leaves the body at death. (A) II, 397; (B) I, 736.

Fohat is connected with Vishnu and Surya in the early character of the former God; for Vishnu is not a high God in the *Rig Veda*. The name Vishnu is from the root *vish*, 'to pervade', and Fohat is called the 'Pervader' and the Manufacturer, because he shapes the atoms from crude material. (A) I, 171; (B) I, 137.

Few word-symbols are more pregnant with real Occult meaning than the Svastika. . . . It is the emblem of the activity of Fohat, of the continual revolution of the 'Wheels'. (A) IV, 158; (B) II, 621.

Of the seven (elements) *— first one manifested, six concealed; two manifested, five concealed; three manifested, four concealed; four produced, three hidden; four and one Tsan* (fraction) *revealed, two and one half concealed; six to be manifested, one laid aside. Lastly, seven small wheels revolving; one giving birth to the other.* Stanza VI, 3.

He builds them in the likeness of older wheels (worlds) placing them on the imperishable centres.

How does Fohat build them? He collects the fiery-dust. He makes balls of fire, runs through them, and round them, infusing life thereinto, then sets them into motion; some one way, some the other way. They are cold, he makes them hot. They are dry, he makes them moist. They shine, he fans and cools them. Thus acts Fohat from one twilight to the other, during seven eternities. Stanza VI, 4.

The process referred to as the 'Small Wheels, one giving birth to the other', takes place in the sixth region from above, and on the plane of the most material world of all in the manifested Kosmos — our terrestrial plane. These 'Seven Wheels' are our Planetary Chain. By 'Wheels' the various spheres and centres of force are generally meant; but in this case they refer to our septenary Ring.

The Worlds are built 'in the likeness of older Wheels' — i.e. of those that had existed in preceding Manvantaras and went into Pralaya. (A) I, 200; (B) I, 168.

The elements of our planet . . . cannot be taken as a standard for comparison with the elements in other worlds. In fact each world has its Fohat, which is omnipresent in its own sphere of action. But there are as many Fohats as there are worlds, each varying in power and degree of manifestation. The individual Fohats make one univeral, collective Fohat — the aspect-entity of the one absolute Non-Entity, which is absolute Be-ness, Sat. 'Millions and billions of worlds are produced at every Manvantara' — it is said. Therefore there must be as many Fohats, whom we consider as conscious and *intelligent* Forces. This, no doubt, to the disgust of scientific minds. Nevertheless the Occultists, who have good reason for it, consider all the forces of Nature as veritable, though supersensuous, states of Matter; and as possible objects of perception to beings endowed with the requisite senses. (A) I, 199; (B) I, 167.

Bear in mind that Fohat, the constructive Force of Cosmic Electricity, is said, metaphorically, to have sprung, like Rudra from the head of Brahma, *'from the Brain of the Father and the Bosom of the Mother'*, and then to have metamorphosed himself into a male and a female, i.e., polarised himself into positive and negative electricity. He has *Seven Sons* who are his *Brothers.* Fohat is forced to be born, time after time, whenever any two of his 'Son-Brothers' indulge in *too close contact*—whether an embrace or a fight. To avoid this, he unites and binds together those of unlike nature, and separates those of similar temperaments. This, as any one can see, relates, of course, to electricity generated by friction, and to the law of attraction between two objects of unlike, and repulsion between those of like polarity. The Seven Son-Brothers, however, represent and personify the seven forms of cosmic magnetism, called in practical Occultism the 'Seven Radicals', whose co-operative and active progeny are, among other energies, Electricity, Magnetism, Sound,

Light, Heat, Cohesion, etc. Occult Science defines all these as super-sensuous effects in their hidden behaviour, and as objective phenomena in the world of sense; the former requiring abnormal faculties to perceive them, the latter cognisable by our ordinary physical senses. (A) I, 201; (B) I, 169.

The Brothers or Sons . . . of Occult parlance (are) the seven primary forces of Electricity, whose purely phenomenal, and hence grossest, effects are alone cognisable by Physicists on the cosmic and especially on the terrestrial plane. These include, among other things, Sound, Light, Colour, etc. (A) II, 278; (B) I, 605.

'The abodes of Fohat are many' — it is said. *'He places his Four Fiery (electro-positive) Sons in the Four Circles';* these circles are the equator, the ecliptic, and the two parallels of declination, or the tropics, to preside over the *climates* of which are placed the Four Mystical Entities. Then again: *'Other Seven (Sons) are commissioned to preside over the seven hot, and seven cold Lokas (the Hells of the orthodox Brahmans) at the two ends of the Egg of Matter (our Earth and its poles).* (A) I, 253; (B) I, 225.

In these volumes it is almost revealed that the 'Sons of Fohat' are the personified Forces known in a general way as Motion, Sound, Heat, Light, Cohesion, Electricity or Electric Fluid, and Nerve-Force or Magnetism. This truth, however, cannot teach the student to attune and moderate the Kundalini of the cosmic plane with the *vital* Kundalini, the Electric Fluid with the Nerve-Force, and unless he does so, he is sure to kill himself; for the one travels at the rate of about 90 feet, and the other at the rate of 115,000 leagues a second. The seven Shaktis respectively called Para Shakti, Jnana Shakti, etc., are synonymous with the 'Sons of Fohat', for they are their female aspects. (A) V, 484; (B) III, 507.

Fohat is everywhere: it runs like a thread through all, and has its own seven divisions. (A) V, 528; (B) III, 555.

It is Fohat who guides the transfer of the principles from one planet to the other, from one star to another child-star.

When a planet dies, its informing principles are transferred to a laya or sleeping centre, with potential but latent energy in it, which is thus awakened into life and begins to form itself into a new sidereal body. (A) I, 202; (B) I, 170.

When Fohat is said to produce Seven Laya Centres, it means that, for formative or creative purposes, the *Great Law*—Theists may call it God—stays, or rather modifies, its perpetual motion on seven invisible points within the area of the Manifested Universe. *'The Great Breath digs through Space seven holes into Laya, to cause them to circumgyrate during Manvantara',* says the Occult Catechism. (A) I, 203; (B) I, 171.

The Spark hangs from the Flame by the finest thread of Fohat. It journeys through the seven worlds of Maya. It stops in the first (kingdom) *and is a metal and a stone; it changes into the second, and behold—a plant; the plant whirls through seven forms and becomes a sacred animal* (the first shadow of the Physical Man).

From the combined attributes of these, Manu, the Thinker, is formed.

Who forms him? The Seven Lives, and the One Life. Who completes him? The Fivefold Lha. And who perfects the last body? Fish, Sin and Soma (the moon). Stanza VII, 5.

The phrase, 'through the Seven Worlds of Maya', refers here to the seven Globes of the Planetary Chain and the seven Rounds, or the forty-nine stations of active existence that are before the 'Spark' or Monad, at the beginning of every Great Life-Cycle, or Manvantara. The 'Thread of Fohat' is the Thread of Life before referred to. (A) I, 283; (B) I, 258.

What is that 'Spark' that 'hangs from the Flame'? It is Jiva, the Monad in conjunction with Manas, or rather its aroma—that which remains from each Personality, when worthy, and hangs from Atma-Buddhi, the Flame, by the Thread of Life. (A) I, 284; (B) I, 259.

PART II

THIS ORDERED UNIVERSE

A Study in Universal Law

Editor
CORONA TREW, Ph.D.

PREFACE

IT IS A TENET of theosophy that we live in an ordered universe, that there is a Plan behind the apparent aimlessness of evolution and that human evolution is an integral part of the whole pattern. This implies that the world, in fact the whole universe, proceeds in its passage through time according to recognizable laws.

Scientific law is more rigorous than the general concept of universal law. The laws of a country can be broken or disobeyed by individuals, they may be changed as circumstances merit such change; but the laws of science are unchangeable. They cannot be broken — they may be disobeyed but in that case the result is inevitable, unavodiable retribution in the absolute operation of the law. A man who 'disobeys' the laws of gravity and steps off a high tower is immediately subjected to those same laws which cause his body to fall at a definable accelerating speed with consequent injury, possible death, of the body. Thus disobedience of scientific law is rather a matter of attempting to ignore the law, or, in other words, ignorance of the law.

Scientific law is thus impersonal and beyond man's control. Theosophy assumes that the laws of the universe are even beyond God's power to alter in the sense that, having set in motion certain clearly defined relationships and rules of behaviour for matter, energy, life and consciousnes, He maintains His universe within the framework of those laws. To do otherwise would be chaos.

The above use of the word law implies that scientific law has a reality of its own. This, of course, is not so. Scientific laws or absolute Laws have no real existence but are mental formulations which express the ordered behaviour of natural phenomena. By means of such intellectual tools man is able to predict what will happen under certain clearly defined conditions and circumstances. Throughout this transaction this meaning should be taken as underlying every use of the word. Natural phenomena are controlled not so much by laws, as by Law, using the word Law in the sense of 'law and order' as opposed to disorder, cosmos instead of chaos.

Man's study of nature, physical and biological, has led to the formulation of laws covering wider and wider spheres of influence. Thus his early attempts merely defined a limited phenomenon as, for example, that all objects fall towards the earth, water flows from hills to valleys. This was later expanded to include the idea of acceleration so that a body falling to the earth has a clearly defined accelerating velocity. The law was then found to have a wider application beyond the earth: the planets, the stars are attracted one to the other by gravitational pull. This was applied to the falling stone by saying that, if the earth attracted the stone, so the stone in its small degree attracted the earth, the two bodies, earth and stone, really moving towards each other. Still more recently the laws of magnetism and rotating bodies have been applied to the idea of gravity with the suggestion that the law of gravity is merely one example of a still wider law.

Thus, in every field, scientific laws are being welded together into more and more widely-embracing theories, the earlier simple formulations being found to be but special cases of the wider law. To some extent it is this idea which has inspired this transaction. A study of theosophical literature and philosophy has led to the view that there are three fundamental activities or qualities of nature, and that ultimately all the known laws of science and of life will be found to be special examples of the three fundamental modes of behaviour. It would be more accurate to say that the smaller laws are special examples of *permutations* of the three fundamental modes since all phenomena fall under the influence of these.

The three fundamental modes of action have been stated in theosophical literature as the laws of evolution, of equilibrium and of periodicity. In human life these may be seen as evolution, karma and reincarnation. This is an oversimplification. This transaction attempts to develop a view of the One-ness of creation, the universality of One Law which we see as Three, and then as many. The reader is asked to study the ideas put forward as broad generalizations. The world is a whole and cannot be dissected without destroying its whole-ness. Thus any correspondences which are given between the Aspects of the Logos, the gunas, the laws of evolution, karma and reincarnation, and laws of physical science must be taken as approximations, since no one law can have one exact correspondence, but must overlap all.

An analogy may serve to illustrate the above remarks. A sheet of paper has a back and a front which can in no circumstances be separated. Let the paper be split to separate back from front. We then have two papers, but each still has a back and a front. Repeat the process, and continue to repeat it until the paper is so thin that the back becomes the front. This can only happen when the paper has no thickness, but then the paper no longer exists, it has disappeared.

That analogy may be taken to a deeper stage of thought by applying it to matter and form. Matter cannot exist without a form. Take a gold watch, destroy the form, melt the gold. You have not destroyed form, but merely a particular form, since the lump of gold still has a shape, a form, if less useful. As matter cannot exist without a form, so one guna cannot act without the others; one of the three fundamental laws cannot work alone and if that is so, then all three must of necessity be expressions of one universal law. That is the thesis of this transaction expressed in different ways by the various contributors.

The Science Group of the Theosophical Research Centre, meeting in London under the Chairmanship of Dr. E. Lester Smith, has worked as a group on the subject of universal law with the object of getting behind the facade of scientific law to the idea of Law itself. The discussions resulted in a number of papers by various members of the group, including one from members working in Manchester, and these papers have been modified by futher group discussions. Finally, Dr. Corona G. Trew has undertaken the onerous task of editing the whole as a composite transaction.

V. Wallace Slater

CHAPTER 6

INTRODUCTORY PRINCIPLES

By Corona Trew

'As above so also below.'

'Diety during its ... Cycles of Rest and Activity, is the *Eternal Perpetual Motion*, the *"Ever-Becoming*, as well as the ever universally Present, and the Ever-Existing". The latter is the root-abstraction; the former is ... a perpetual, never-ceasing evolution, circling back in its incessant progress through aeons of duration into its original status—*Absolute Unity'*.

The Secret Doctrine, IV, 115.

THE SECRET DOCTRINE PRESENTS the student with a magnificent picture of Logos within Cosmos, and of Man within the Universe, each unfolding spiritual powers in accordance with certain principles of cosmic evolution. In the Proem we find a statement of three universal principles which are basic to all manifestation whether it be that of a universe, a world, a man or an atom. One may regard these principles as one expression of that great triple rhythm of creation into which all the laws of nature can ultimately be resolved.

The Proem first postulates the existence of an ultimate principle of spiritual Being, out of which all comes, and to which all ultimately returns bearing the fruit of the evolutionary process. This original principle is omnipresent, eternal, boundless, representing Be-ness rather than Being, and is the parent of Cosmic Ideation as well as of Cosmic Substance, that is, of both spirit and matter. In man it is the source of both consciousness and of its vehicles.

In the second phase, this primary principle becomes active as eternal, perpetual motion and manifests in a multiple field of being and becoming. The field is our universe, in which the great rhythm of periodicity reigns, described as a universal law of 'periodicity, of flux and reflux, ebb and flow, which physical science has observed and recorded in all departments of Nature'. *(The Secret Doctrine,* I, 82.*)*

From primary Being the creative process emerges as an impulse outward into action and motion: the latent potency moves into an active phase. The very tension, set up by this activity driving outwards from its centre of rest produces a restoring force, which ultimately draws the creative process back towards that centre. Hence arises the periodic rhythm of the evolutionary process, with its various phases. The tendency ever to rest in Being is succeeded by the tendency to move into 'becoming' and the flow again towards Being causes a 'having been', and so the cycle passes back to Being again. Thus is set up an eternal triple tide, one of the fundamental movements of the universal life. Within this triple movement takes place all the processes now known under the term 'evolution'.

Life and consciousness emerge within this complex field, and the third great principle of *The Secret Doctrine* affirms what is termed the 'Cycle of Incarnation or Necessity'. Here, life passes into a recurring cycle of necessity. Through incarnation in form—of atom, cell, plant and animal—life increasingly expresses its potentialities. Finally emerging as man it begins a new pilgrimage, that of the soul through greater and lesser cycles, by successive incarnations that take him through races, rounds and Chains. Under bondage to law on the outward sweep of the path, on the return cycle of the great breath, commonly termed the path of return, man gains freedom to work within the Law he now understands and at the end becomes a master of Law. He arrives at understanding by a conscious identification of himself with the interior or subtler levels of awareness. When human consciousness, thus unfolding from life to life, has won a measure of self-realization by mastering the forms through which it passes, it can become infinitely free by working with the law that ever 'makes for righteousness'.

In such manner universal law is said to reign throughout the Cosmos and all that manifests within its fields expresses the triple rhythm of progression from Being to becoming, then to 'having been' and thus back again to Being at a deeper level of realization, whence begins afresh a new cycle of experience. Life in the kingdoms of nature is under bondage to this triple rhythm, but consciousness in man may win release to use and control the law, through realizing itself as one with the source of all, and hence itself creative. This is true of man alone, a privilege not shared with the other orders of Nature.

The following essays are a study of the working of these principles in their various aspects and as they are expressed in nature and in man.

CHAPTER 7

THE TRIPLE BASIS OF LAW
By Corona Trew

'Let the student remember that number underlies form, and number guides sound. Number lies at the root of the manifested Universe: numbers and harmonious proportions guide the first differentiations of homogeneous substance into heterogeneous elements, and number and numbers set limits to the formative hand of Nature.'
The Secret Doctrine, V, 418.

'Laws of science have no ultimate reality in themselves, but merely express principles which, behind the world of form, energize and maintain an ordered world.'
Study Course 'Universal Life & Law,' 1950.

'We recognize but one law in the Universe, the law of harmony, of *perfect* EQUILIBRIUM.'
The Mahatma Letters to A.P. Sinnett, 2nd Ed., page 141.

UNITY AND THE EMERGENCE OF TRIPLICITY

THE SCIENTIST AND THE OCCULTIST approach the search for the laws of nature from different viewpoints. A scientific law may be defined as 'a statement or formula expressing the constant order of certain phenomena'. Since knowledge is progressive, scientific law is subject to revision in the light of further investigation and experiment, and so the scientific attitude to law is that of progressive discovery of relations between phenomena.

From the point of view of the occultist the universe is seen as emerging from a 'within' to a 'without', or less correctly expressed, from an 'above' to a 'below'. The fundamental occult attitude to law may be summed up in the phrase 'as above, therefore, so below'. This statement should not, however, be transposed into the common misapprehension 'as below, so above'. Natural law according to the occultist arises from the tendency of life to take on certain patternings from within in response to the urge of the life-force, known as Fohat, and thus to specialize itself or even to crystallize along certain lines. From the viewpoint of the occult scientist, the creative principle within the universe, that has here been termed the Logos, has fundamental characteristics, and as that Logos moves outward into manifestation, reflecting Himself in his objective universe, the qualities that are characteristic of the Logoic nature arise in that universe. Natural laws thus express a fundamental interaction of the Logoic spirit (Life) with matter (the field of manifestation), and the qualities of both spirit and matter determine the nature of law.

The Logos of our System, possessing certain inherent characteristics, is probably restricted by having to work within the framework or terms of reference set by the nature of the matter of his particular field of activity. In this sense He is conditioned by the external, Cosmic laws. What is termed occult law is, for our universe, the resultant in the universal mind (Mahat) of that conditioning. Occult law expresses the inter-action of spirit and matter, of life and form in any given field. It is in this sense that the statement that there is but one law in the universe, that of harmony or perfect equilibrium, may be interpreted.[1]

---

[1] *It is within the illumined mind of man that the opposing viewpoints of scientist and occultist may be resolved. The scientific mind sees law from below, from the outside, and climbs painfully towards a synthesis by experiment and observation and reasoning. The occultist sees law from within as the expression of living principles, but perhaps loses detail in the vision of the whole. Within the illumined mind of truly human man the scientific and occult viewponts are experienced as complementary aspects of one whole, the universal and absolute law of harmony and rhythm. Deduction and induction are both necessary if man is to know the essential nature of himself and the Cosmos.*

So close is the interaction between the creative Logos and his universe that we cannot think of manifestation in simpler terms than of this interweaving. In its simplest aspect the universe expresses wholeness, or unity, because it is the expression of a single creative consciousness. The unitary principle is the fundamental one from which all other law proceeds. It is at the root of, and precedes, all things, and as a result of the underlying unity, all differentiated beings strive to return to wholeness and harmony.

The purposeful interaction of a creative principle with material substance is the background for the universal law of harmony. Within this harmony certain fundamental relationships may be perceived. These relationships apply to both the universe and to man, for the ancient wisdom has always taught that man the microcosm is a miniature replica of the whole macrocosm, the universe, and shares its essential nature and patterning. As the galaxy or island universe, within which we live, is a unit in space, separate from, yet coexisting with countless other similar universes within the Cosmos, so each man is an organic unity, and remains a true whole in himself, related to other men in the larger organized unit of society.

Every whole is a completely self-contained unit, which at its own level is endlessly extensive and without limit for that which has experience within it. Its limiting boundary or ring-pass-not only emerges as a fact of experience when life or consciousness seeks to step out from the level or standpoint of that unit into a larger and more inclusive sphere of activity.

In terms of mathematics all numbers and their relationships may be resolved into that whole which is represented as a point or centre in a circle. Conversely, from the movement of the point within its boundary circle the whole system of numbers may be unfolded, as was shown in the ancient systems of the Pythagoreans and other similar schools. Indeed the whole Cosmos may be expressed in terms of this symbol. (*The Secret Doctrine*, IV, 115.)

At the physical level the atom is the real unit or whole upon which the dense material world is built; the cell, from which all plant organisms simple and complex are built, forms the unit of the vegetable kingdom. The animal form, and its associated group-soul,[2] represents a further unit, whose periphery limits intensively the consciousness that uses it. It is only man with his wider vision who can perceive the limitations of these wholes and see that they are elements in a larger unity. For man, the unit of consciousness is the individualized human spirit, using as its personal medium of expression mind, emotions and physical body, the so-called personality of man. The harmonious expression of his spiritual consciousness in the field of organized society is the human task, which involves the achivement on his part of an integrated means of expression in that field. The integration of personality is only possible because the true human unit lies outside the organism that has to be integrated, at the causal or egoic level. When a man learns to live consciously in his own nature at that interior level, he can direct personal behavior wisely according to the deeper laws. One of the more important of these is the law of brotherhood. By living according to that law some measure of harmony in the field of social relationships is achieved and a yet wider vista opens up before him. He may then contact the larger whole of the universal life and finally become a channel for the spiritual (atmic) force that is the basic power of the whole system.

Thus the first law of manifestation, the unitary principle of wholeness, or one-ness, has a valid meaning at every level, and unity of organization and form is linked with unity of purpose or life-function for every organism. It follows that psychological wholeness and harmony is a necessary development for humanity, just as unity of function at the biological level is necessary for the living organism.

---

[2] See Annie Besant, *A Study in Consciousness*, 87-89.

The development of polarity within the original unitary condition leads to the manifested universe, the product of interaction between a creative principle and material substance. Manifestation as we understand it emerges when spirit and matter begin to interact, forming the eternal poles upon which the universe is spread. In eastern literature the Self and the Not-Self and their interaction are said to lead to manifestation in form. This polarity is the *second phase* in every manifestation. There is not only a duality of action and reaction at each level, but there is a polar interaction between one level and another. The number of levels involved depends upon the elaboration of the organism. Human consciousness, for example, can range over three levels of experience, at the least, namely spiritual, intellectual and sensory, but in all experience a fundamental subjective-objective relationship is maintained. Within any complex organism there is always this polarity which lies in the interplay of experience between any two levels, one of which remains subjective while the other is in the realm of objective form. These two represent the opposite poles of experience for the consciousness working in that organism.

In the universe as a whole, the fundamental polarity lies between universal mind, Mahat, and Cosmic Substance. All the way out and down it is expressed in the subject-object relationship. In man it shows in the two chief levels of awareness termed the subjective and the objective, which more specifically may be called the operator and the instrument of action. Everyone can become aware of this duality of his being, although the level of the focal point at which one shifts into the other varies considerably for different individuals, and at different times in the same individual.

> 'We find first of all two distinct beings in man; the spiritual and the physical, the man who thinks and the man who records as much of these thoughts as he is able to assimilate.'
>
> H. P. Blavatsky, *The Key to Theosophy*,
> Third Edition, 62.

The recognition of polarity is found in the structure of language itself, the medium of communication between man and man. The simplest sentence must have subject and object, linked by an operative relation between them, the verb, and out of this relationship a sentence is formed. Thus, for example, 'John' as subject and 'the ball' as object have no relationship unless linked by some verb of action such as 'hits', 'drops' or 'catches', and graphic situations may be evoked if the verb is 'steals', 'slashes' or even 'swallows'. Two smaller wholes, subject and object, when brought into a dynamic relationship express a larger unity. In the process of unfoldment of the Cosmos, the latent Deity and the watery deep of matter remain aloof and apart from one another until the divine action stirs the waters into motion. The spirit of God *moved* upon the face of the waters and the whole creation was brought to birth. This representation of the universe as a harmony of interaction between two complementary principles synthesized in a deeper unity, which sustains both, is expressed in the Proem of *The Secret Doctrine*.

> 'The Manifested Universe, is pervaded by duality . . . but the opposite poles of Subject and Object, Spirit and Matter, are but aspects of the One Unity in which they are synthesized. . . . There is "that" which links Spirit to Matter, Subject to Object. This something . . . is called by Occultists Fohat. It is the "bridge" by which the Ideas existing in the Divine Thought are impressed on Cosmic Substance as the "Laws of Nature".'
> *The Secret Doctrine*, I, 81.

THE TRIPLE BASIS OF LAW

The laws of nature then, are the result of the interaction of a spiritual principle, expressing itself in the mental world of our system, and the universal principle of form or substance which we recognize as matter. The polarity of these two opposites sets up a tension which is resolved by their interaction, and this interplay constitutes a third principle. The universe, therefore, expresses a triplicity in all its aspects and at all its levels. This triplicity may be detected in cross-sections of the whole cosmos, from the 'within' to the 'without', in familiar trinities, some of which are given in the following table.

Table I

TRINITIES IN NATURE

Subjective Pole	Bridge	Objective Pole
Spirit	Fohat	Matter
Cosmic Ideation	Cosmic Mind (Mahat)	Cosmic Substance
Logos	Man	Cosmos
Consciousness	Self-Consciousness	Vehicle or 'Body'
Self	I	Not-Self
Monad	Ego	Personality
Spirit	Soul	Body

In every case the middle member of each of these trinities owes its existence to the interaction of the two extremes, and yet paradoxically without it there would be no manifestation. Arising from the interaction of the two, it yet supplies the essential relationship that makes of the three a functional unity. Of the various trinities given in the Table, that of Logos, Man and Cosmos is of supreme importance to us, for Man is the 'bridge' which unites Logos and Cosmos, and within him is the synthesis of all opposites. Logos and Macrocosm meet in Man the Microcosm.

Fundamentally for humanity there are three great regions of experience, as *The Secret Doctrine* shows. These may be termed spiritual, intellectual, and material or sensory, and within each of these again the law of harmony is expressed in a triple rhythm. The middle member of these three great streams of evolution, the intellectual, is the linking bridge which, like the verb in a sentence, brings the interplay of the other two into active manifestation.

> ... there exists in Nature a triple evolutionary scheme... or rather three separate schemes of evolution, which in our system are inextricably interwoven and interblended at every point. These are the Monadic (or Spiritual), the Intellectual, and the Physical Evolutions. These three are the finite aspects, or the reflections ... of Atma ... the ONE REALITY.' 'Each of these three systems has its own laws . . . Each is represented in the constitution of Man, the Microcosm of the great Macrocosm, and it is the union of these three streams in him, which makes him the complex being he now is.'
> *The Secret Doctrine*, I, 233.

There is also a triple rhythm discernible in the nature of matter itself at every level. At each level the creative impulse manifests as three aspects. The first is stabilizing, the second harmonizing and the third represents an active energizing agency. These three arise in manifestation from that same interaction of the creative principle with material substance which gives rise to the triplicities indicated in the previous table. Something of this relationship is expressed in the following table. The sub-division of the intellectual 'stream of evolution' into two parts arises from its nature as a median or middle term between two poles. It may be polarized more specifically to the spiritual realm giving the higher ranges of intellectual faculty and powers, or it may be polarized towards the more material, sensory or physical pole of the manifested universe leading to the more objective and concrete lower mental faculties and activities.

> ' . . . everyone will agree that the intelligence of man is *dual*, to say the least; e.g. the high-minded man can hardly become low-minded; the very intellectual and spiritual-minded man is separated by an abyss from the obtuse, dull, and material, if not animal-minded, man.... Every man has these two principles in him, one more active than the other ... These, then, are what we call the two principles or aspects of *Manas*, the higher and the

lower; the former, the higher Manas, or the thinking, conscious Ego gravitating toward the spiritual Soul (Buddhi); and the latter, or its instinctual principle, attracted to *Kama*, the seat of animal desires and passions in man.'

The Key to Theosophy, Second Edition, p.120

Table II

THE TRIPLE RHYTHM IN THE THREE REGIONS OF THE UNIVERSE

Evolutionary Region or Level	First Aspect Stabilizing	Second Aspect Harmonizing	Third Aspect Energizing
Spiritual	Will	Being	Creative activity
Intellectual (a)	Volition	Love-Wisdom	Ideation
Intellectual (b)	Desire	Feeling	Thought
Physical	Inertia (Tamas)	Rhythm (Sattva)	Mobility (Rajas)

While all three elements of the triple rhythm manifest at every level, at each level one tends to be more dominant than another. Thus, the harmonizing principle reigns predominantly in the spiritual world, the energizing in the intellectual and the stabilizing in the material. The spiritual world is dominantly one of potent being and harmonious relationship, the intellectual is par excellence a world of active mobility and in the physical world the quality of inertia and solidity is more marked than in either of the other two.

Something of the influence of these elements within the personal field of man may be conveyed by a further extension of the analogy of language, for we may express the basic attributes of these three elements acting in the levels of experience by the following active states, defined by a subject and transitive verb.

Table III

Region	FIRST Aspect	SECOND Aspect	THIRD Aspect
Spiritual	I will	I am	I bless
Intellectual (a)	I wish	I understand	I design (plan)
Intellectual (b)	I want	I love	I think
Physical	I hold	I sense	I act

THE NATURE OF THE THREE QUALITIES

An understanding of the nature of these three fundamental qualities is basic to an occult study of the laws of physical nature, for all natural law is held to be the result of their interweaving. They may be regarded as fundamental 'modes of motion' set up by the interaction of the spiritual and material principles. The three Sanskrit words *tamas, sattva* and *rajas* are the names given to these principles or qualities which are termed 'gunas' in Sanskrit philosophy. They are usually translated inertia, rhythm and mobility.

The quality of tamas or inertia is that property which manifests in a centripetal manner. It holds and attracts to a centre, restores to rest or to the original focus of action. It gives coherence such as is found when the drops of a liquid tend to coalesce together. It gives to matter its inertia, position, stability and mass, and since it is the principle which predominates in matter and the physical pole of the universe, it gives rise to the gravitational forces which tend to draw matter down to its lowest point. It provides the gravitational tendency of the cosmos, that we recognize as the mass, or inertia, of matter. In relation to light it is the principle of 'darkness', produced by the total absorption of light by the body upon which it falls. The theoretical 'black body' of radiation theory is the complete expression of unrelieved tamas. In the properties of a body at rest, tamas gives it inertial mass, while within a body such as an atom, tamas is the centripetal force tending to draw the peripheral material ever towards the centre. When a body is travelling along a straight line the tamasic quality is found within the momentum of the body as the *mass* component of the momentum, that component which contributes most to the capacity for doing work, or for massive destruction, of the moving body. For example, when a body is moving in a circular path the tamasic element is, as indicated above, the force which draws it in to the centre of its circular path or orbit.

At the opposite extreme the rajasic quality is that which ever thrusts into action onwards and outwards. It tends to disruption and dispersal of matter, and in a liquid tends to cause it to break down into differentiated droplets. It is especially associated with electricity and its action is noticeable in the tendency of an electrically charged liquid to break down into microscopic droplets of charged 'fog'. It thus results in an outward-flowing mobility in nature. Within a resting spherical body, such as an atom, although the tamasic quality appears to predominate, yet rajas is found in the electrical content or charge of the body: the peripheral electrons moving in their 'orbitals' by the force of rajas. This force acts tangentially to the orbit as a centrifugal force, being balanced by tamas, the restoring centripetal force. This is true both of an atom, and in a body travelling around a circular path. For a body travelling in a straight path rajas is the linear velocity-component of the momentum (momentum = mass x velocity) which gives to the static mass or inertia its driving force.

The middle principle, sattva, is both the balance between these two and the resultant of their interaction, and is dual in its mode of behaviour. It gives balance, equilibrium, rhythm and harmony and as the third or linking principle is essential to the existence of matter. Within an atom the balance between the inertial mass, the centripetal force of tamas, and the distributive electric charge of rajas acting centrifugally, gives to the atom its existence as an entity. Sattva gives the cyclic form of the atomic and planetary orbits when rajas and tamas are exactly balanced in a closed system. The precessional motion which converts a closed cycle into a spiral path, results when the rajasic guna is applied in a different dimension and so disturbs the sattvic rhythm which in its completeness would lead to a static and perhaps stagnant state. In terms of light, while rajas gives the luminosity of a body which shines by *reflection*, a sattvic body is self-luminous. There is also a higher octave of sattva. In a special way it is the expression of the universal law of harmony and wholeness, which is dealt with more fully in a subsequent chapter.

It is said that the 'Qualities' move amid the 'qualities'. This refers to the guna-principles in their transcendent and immanent aspects respectively, for their manifestation at a subtler level will influence that at any exterior level. The three gunas, or Qualities, can also be detected in the laws that apply to various levels of experience as well as in the physical laws of nature. The great laws of human growth, of human evolution, including reincarnation and karma, are also expressions of the threefold play of the qualities called the gunas.

CHAPTER 8

UNIVERSAL LAW IN NATURE
PERIODICITY, EVOLUTION AND KARMA
By CORONA TREW and DOROTHY ASHTON

'Under the contemplative gaze of Consciousness, three tendencies manifest themselves within the Matrix. . . .

'These three *gunas, sattva, rajas* and *tamas,* are the strands of which the twisted rope of being is woven. All things from grossest "matter" to subtlest cosmic thought-stuff, are the manifestations of one or more of these three tendencies, and it is one of the tasks of the disciple to analyse all phenomena in terms of these *gunas.* His effort is to be able to stand firm in *sattva....* He must therefore be able to say of any phenomenon: "this is *sattvik* for it brings increase of Light and harmony and so will lead me upwards; this is *rajasik* for it leads but to motion and is founded on desire; this is *tamasik* for it fills the soul with darkness, taking it captive to an outer Fate".'

Sri Krishna Prem,
Yoga of the Bhagavad Gita, 135, 139.

THE TRIPLE RHYTHM of creation, that gives rise to the three gunas or qualities, expresses itself in nature in three fundamental modes of manifestation. These, like the three Logoic aspects with which they are correlated, are all in action at once, one or another merely predominating in any given phenomenon. They may be traced at all levels throughout the whole system. In attempting to give names to the laws that arise from these fundamental modes, the first aspect appears always to impose limitation and what may be termed sacrifice, the second is productive of equilibrium and harmony, while the third is expressed in energy and action. Sacrifice, harmony and activity may be said to characterize the qualities in nature.

The first or stabilizing aspect of the creative process, that known as tamas guna, expresses itself everywhere in terms of limitation and of sacrifice; it governs the incarnation of life in form. It gives rise at each level to a certain limitation, because it is a force that establishes a field of activity by defining a ring-pass-not. Such a boundary must be established before any manifestation can take place, whether in the cosmos as a whole or in any body, organism or atom. The limitation is not merely in terms of space — the isolating of a portion of material within which to operate — it implies also privation in terms of relationship and expression.

> 'The sacrifice imposed is a deliberate withdrawal from a larger field in order for a while to concentrate on a comparatively narrow issue. . . . We may glimpse something of the real meaning and joy of sacrifice, for, in terms of consciousness, success in a given cycle means that a certain maturity has been attained and that a ring-pass-not may then be broken and a field with a wider horizon entered and explored.'
> E. L. Gardner, *The Ring-Pass-Not and Sacrifice.*
> *Theosophy in Action*, June 1951.

The characteristics of tamas guna have been described as follows:

> 'it is that aspect of the One which limits and restrains, offers resistance, stabilizes and is somnolent. . . . It is under this aspect that the world of objects arises descending towards the objective physical plane, from chaos.'
> Corona Trew
> *Studies in the Secret Doctrine*, Study IV.

The way in which this law of limitation manifests in the three great regions of experience may be seen in Table II of the previous chapter. At the spiritual level the quality of tamas, of limitation, is expressed in the act of incarnation. Pure will (atma), a free creative energy, becomes identified or clothed with a denser form. The heavier, more constricted material limits the field of expression of power, but the power gains thereby in potency. Electricity affords a useful illustration of this law at work. When electricity moves through the wires of a variety of electrical appliances, it is harnessed and provides a source of power that has changed mankind's whole manner of life. Within the intellectual field the tamasic quality of the forms so readily built, at first satisfies then fails to meet man's hunger for understanding. Thought constricts and defines activity, and the 'charge' thus built up drives the seeker after ultimate truth to deeper, more penetrative, intellectual activity.

In the lower regions of the intellectual field, under the sway of the kamic forces, tamas manifests as desire, that driving force of primitive humanity and of the animal kingdom. Here it has been described as 'will dethroned'. At the physical level it gives the quality of inertia and of solidity as discussed in the previous chapter.

The middle, or harmonizing quality, sattva guna, is best studied after considering the energizing aspect of the divine purpose. This third aspect, rajas guna, is in many ways the antithesis of the first. It expresses itself in the laws of activity and of motion. It tends to drive outwards into space dispersing energy from the centre of any form created by the first aspect. In the complex process of evolution it is expressed in the outward drive of life tending to seek experience through new, and ever more complex, sequences of forms. It is the rajasic element in the cosmos that gives rise to that avidity for experience that leads life to experiment with, and to organize, a multiplicity of forms. As an outward and onward driving energy it produces change and governs evolution as a sequence of forms in progression.

> 'It is vigorous; it expands and breaks limits; it is impulsive, ardent, has appetite, energy. Under this

> aspect the One differentiates the various "living" forms which have senses to contact the objects of the senses projected under the tamas-guna.'
> *Studies in The Secret Doctrine*, Study IV

Like tamas guna, its manifestation may be observed at every level of the universe. At the spiritual level it is the principle of pure creative activity, the essence of all that we know as creation: the Spirit of God that moved over the face of the waters, and caused Light and the universe to be. At the upper intellectual region it may be termed ideation, for here it contributes the human faculty of planning and the multiplicity of archetypes and ideas of which all lesser forms are expressions. In the lower form-building levels of the mind it gives rise to the process of thought association and concrete form building. At the physical level it is the active principle of motion, leading life to evolve through response to external stimuli. All the forms of the various kingdoms are used and discarded and left to die out when the life moves on into other more resilient forms capable of expressing a deeper range of its powers. In this way life 'evolves' the forms, and evokes its own latent powers.

Hence, while tamas guna causes the limitation of life in forms and thus makes for sacrifice, and, in a sense, privation, rajas guna drives these forms into activity, division and differentiation. 'The mind is hard to curb and restless as the wind', according to the Bhagavad Gita, because the mind especially embodies the quality of rajas. Hence the rajasic quality dominates the middle region of the universe, and the intellectual phase of the evolutionary cycle.

Holding the balance between rajas and tamas is sattva guna, which tends to restore equilibrium between the opposing tendencies of the other two. Its tendency is to resolve all things back into a basic relationship of harmony. It gives rise to balance, to the relationship of action and reaction, of inbreathing and outbreathing, and expresses the urge to return to the underlying condition of harmony and equilibrium.

Like the other two, the harmonizing sattvic principle is present in the whole field of experience. In the realm of the spirit it is the quality which may be termed *Being*, the balanced harmony of life residing in itself as a perfected whole. In the higher realm of the dual intellectual field it is the quality of love-wisdom, the pure and essential irradiation that characterizes illumined thought. At the lower mental level it may be seen as feeling; the indwelling of life within the substance of the lower part of the intellectual realm, that gives it a vital immediacy and something of the unifying quality associated with water or any similar fluid. In the physical world it is expressed, as shown in Chapter I, as a rhythmic balancing and equilibrising force.

THE LAWS OF EVOLUTION AND KARMA

The harmonizing or sattvic element in nature is, however, more than just the restorer of a balance between the other two. It is said in *The Mahatma Letters* that 'We recognize but one law in the Universe, the law of perfect equilibrium' (*page* 141). The fact that the universe emerges from a harmonized state into one of progressive cyclic movement implies that behind and within the interplay of the three guna forces there lies another range of influences, and that it *was from a sattvic condition that our system emerged.* It thus has a sattvic 'root' from which it has developed and to which it will return the fruits of its adventure. Moreover, at any point of assimilation from which a fresh effort emerges, the temporary dominance of sattva guna leads to a 'fourth' mode of manifestation called the 'turiya' state, a point of balance that synthesizes the three forces and sums up all that has gone before.

> 'Entering the Turiya state may be described as approaching the maturity of perfection — a critical stage of suspense before proceeding further.'
> E. L. Gardner, *The Heavenly Man*, 15.

The complete rhythm of growth by cycles, and hence the process of evolution in its fullest sense, is thus an expression of the interaction of all three of the basic qualities or modes. In the drive down into form known as involution, the tamasic element is dominant, tending to cause the limitation and shutting off of the powers of life — so that life is then said to be 'crystallized and immetalized in form'. In the phase to which the term evolution is generally given, in which we find the vast multiplication and differentiation of forms, the outward rajasic thrust is dominant. In the third phase, that of the spiritual evlution of the individualized life, sattva guna gradually takes control. In the complete involution-evolution process, taken as a whole, there is a release of the latent powers and faculties of life into ever fuller expresion. The sattvic principle, when dominant in the final stage, holds the three in a stable equilibrium that allows the system to return to poise, or harmony, with its achievements assured.

Thus, seen as a whole, the process of evolution, through incarnation in form, from the simple to the complex, from atom to cell, mineral to plant, and animal to man, is the expression of all three of the qualities, and is a spiral process rather than a closed cycle. Whenever sattva guna restores harmony between the tamasic and rajasic qualities it leads, not to an empty return to the *status quo ante*, but to a new point of balance, the 'fourth' mode, and such is the operation of sattva-guna that the ensuing cycle opens at a higher octave.

The Secret Doctrine refers to this process as an expression of the law of Occult Dynamics; that *'a given amount of energy expended on the spiritual . . . plane is productive of far greater results than the same amount expended on the physical objective plane of existence'.* (*The Secret Doctrine*, II, 369.) Sattva guna, dominant at the spiritual level, has in its operations in the lower worlds sufficient reserve power at each point of change to lead the evolutionary process a stage nearer to fulfilment.

Thus, whenever life moves from one octave to another, from expression in simple to more complex forms, from a lower kingdom to the next higher, it is due to the operation of this higher sattvic potency. Restoring balance within the lower cycle, sattva releases more of the latent power of life, and subsequently of consciousness, to further experience and expression. The whole process of evolution is thus like that by which the tide comes in to the sea shore; progressive waves, lesser and greater, gradually move inwards and press on up the beach as the tide flows in with irresistible force. Each cyclic experience, each incarnation in form, and—on the larger scale—each round and Chain, is unique, like a single wavelet or wave, but, under the ever present harmonizing influence of sattva life tends spiritually forward and moves towards its destined purpose, much as the tide flows in.

In the long sequence of Chains, rounds and cycles, the life-wave goes through all the kingdoms from simple to complex, ever enriching its focal centre of experience until the complex being known as man emerges upon the evolutionary scene. He, again, in his earlier cycles of experience, is repeatedly carried downwards and outwards from his true spiritual home into experience in physical form under the dominating sway of the tamasic and rajasic forces. At this stage, stimulated by the qualities of darkness and activity, he proceeds upon what has been termed the 'Pravritti Marga', the human path of outgoing.

> 'Two mighty tidal urges rule the worlds, and both of them are living spiritual powers. One is . . . the great outgoing Creative Breath by which not only is the universe spread forth in space, but all the inner life of thought and feeling flows outward seeking whom it may devour. This is the urge of self-assertion, self-expansion, survival of the fittest. . . . Here is the inner cause of war and all the selfish life of competition. . . here as well, the force behind man's mind, wheeling in ever-widening circles to receding frontiers.'
>
> Sri Krishna Prem,
> *Yoga of the Bhagavad Gita*, 108.

Thus, on the path of outgoing, the two qualities, tamas guna and rajas guna, dominate the latent human spirit, but at the turn of the cycle the restoring force of harmony slowly and irresistibly comes into play, and the forces ruling the path of outgoing slowly yield to the sattvic influences of the path of return.

> 'The second movement ... is the *"nivritti"* or Homeward-flowing Tide. By this all the rich treasure of experience, the Fruits of the World Tree, are gathered in once more to the One Life like mighty rivers flowing homewards to the sea.'
>
> *Prem, ibid,* 108.

Such is the occult view of evolution as a whole. It is seen as a vast interwoven and cyclic activity of three major forces, acting rhythmically in relation to each other as living, active potencies. 'The end is consummation sweet', the assimilation of experience, and an ultimate readiness for further adventure.

The Secret Doctrine not only views the law of evolution in this deeper light, but also asserts that the law of karma is an expression of the working of the higher or cosmic sattva. Karma is that one law of harmony already referred to. The action and reaction of karmic forces helps to resolve the stresses and strains induced by the other two modes into further harmony.

> 'The only decree of Karma — an eternal and immutable decree — is absolute Harmony in the world of Matter as it is in the world of Spirit. It is not, therefore, Karma that rewards or punishes, but it is we who reward or punish ourselves, according as we work with, through and along with Nature, abiding by the laws on which that harmony depends, or — breaking them. ... If one breaks the laws of Harmony ... "the laws of life" ... one must be prepared to fall into the chaos oneself has produced.'
>
> *The Secret Doctrine,* II, 368.

Such a breaking of the 'laws of life' is the fate of man only upon the outward going path under the sway of the forces of darkness and activity. Once the path of return is entered upon and the sattva guna reasserts itself, the expenditure of even a small amount of energy at the spiritual level will produce cumulative results and the universal harmony will once more reassert its sway.

The influence of the higher or cosmic sattva is the cause of the acceleration principle in spiritual evolution. Once a man has gained a glimpse of his spiritual goal and presses on steadily towards it, in spite of difficulty and failure in the earlier efforts there comes a time when a rapid expansion of his faculties takes place. The true sattvic nature of the Self begins to express itself and thenceforward his progress is as rapid as the unfoldment of a flower meeting the warmth of an early spring day, a visible expression of the sattvic harmonies.

On the human path of outgoing the embryo human spirit does not know its powers or its own nature, dominated as it is by the forces of tamas and rajas; of darkness or constriction and the drive towards activity. It makes its adventures largely automatically under the influence of these 'blind' and delusive influences. But the human spirit itself is dominantly sattvic, and cosmic sattva, dominant within the human ego, acts to bring about a restoration of harmony. This action-to-restore-harmony is known as the law of karma. It is the reaction of cosmic sattva to the disturbance of its equilibrium produced by the earlier action of tamas and rajas. Such reaction is automatically and entirely just: essentially educative and purgatorial, not punitive. It entails suffering only on those who resist its action. Even so they will become strong in resistance, extracting the lessons and power of tamas, or highly skilled in evasive activities, learning the lessons of rajas. In the end they must learn also the art of harmonizing opposites within themselves, because the human spirit has the sattvic quality as its dominant characteristic.

CHAPTER 9

UNIVERSAL LAW IN NATURAL FORMS AND ORGANISMS

By Corona Trew and Dorothy Ashton

'Our conception of wholeness must include time and duration, for every part of Nature is history and change. Only the guiding energy field of the whole is ultimate and persistent. The Genius of the Whole is the highest level of the Cosmos. It is not an emergent deity in the sense that it is dependent on lower levels. It permeates all that is in space and time and is the guiding energy field on which all order depends. The Whole as such does not evolve. Evolution is a local process in which the environment furnishes the original stimulus. Matter and spirit are two fundamental elements in a hierarchy of fields. The structure, rationality and beauty of the lower levels imply not a blind impetus from below, but an organizing formative stimulus from above.'

Bernhard Mollenhauer,
Horizons of the Occidental Mind,
Hibbert Journal, Jan. 1952, 171.

'The structure of value is eternal, but our participation in it is in our own control. To choose freely to realize in our individual lives the beauty of spirit is the great affirmative; to refuse to do so is the great negative.'

John Booden, Nature and Reason,
Contemporary American Philosophy, Vol. I.

In the preceding chapters three great tendencies have been traced throughout the regions of the universe, as active forces or qualities, imposing tidal rhythms upon all life within that universe. The further expression of these qualities within living organisms, human, animal and plant, and in mineral forms, gives rise to the laws governing the forms of Nature's kingdoms. Here we may observe something more nearly approaching scientific laws.

The three main regions of the universe, the spiritual, the dual intellectual and the material, correspond to centres of awareness for the various living organisms which inhabit that universe. Any entity that exercises self-conscious choice possesses a specific focus of life at the spiritual level, which expresses itself in relation to the lower worlds as awareness of experience. Impulses from such a spiritual centre or focus, impinging upon the external material world, are gradually discerned, then fully experienced within the field of the mind. Ultimately this gives birth to that precise awareness known as self-consciousness, which is able to discern relationships between the within and the without. The process may be compared to the brilliant light of an electric arc which is generated in the gap between the positive and negative poles of an arc lamp.

> 'Apart from Cosmic Substance, Cosmic Ideation could not manifest as individual Consciousness, since it is only through a vehicle . . . of matter that consciousness wells up as "I am I", a physical basis being necessary to focus a Ray of the Universal Mind at a certain stage of complexity. Again, apart from Cosmic Ideation, Cosmic Substance would remain an empty abstraction, and no emergence of consciousness could ensue.'
> *The Secret Doctrine*, I, 81.

The energizing force behind this self-consciousness is, however, impersonal and universal. It is the force known as 'Fohat' in *The Secret Doctrine*, and its particular impact at the spiritual level is termed Fohat acting as the 'light of consciousness'. Although spiritual self-awareness is only fully unfolded when man, triumphant and illumined, consciously enters the realm of a true Logos or creator of a universe, it begins to dawn within the human kingdom. From the interaction of human volition and material substance within the field of man, there arises—or can arise—a gradual domination of matter and mind by spirit. This, as indicated at the close of the previous chapter, occurs on the path of return, as sattva guna comes more fully into play and leads evolution to its climax.

Within the spiritual region are expressed the occult laws of creative self-awareness. Those entities who have learned to work consciously at this level are free, that is they are in command of the forms and forces of the denser worlds. They understand the transmutation of form through the alteration of interior impulses. They possess also the ability to create a form at will to serve the purpose of life in any lower region.

Within the intellectual region of the universe there are also focal centres through which life flows into and acts upon form and presses thence into objectivity. As shown in the previous chapter the intellectual field of expression is dual and life builds within it two kinds of centres for experience in form. In the subtler mental region life requires a focus to serve both as a permanent nucleus from which to thrust outward into the intellectual and denser material fields of existence, and as a repository for the essence of the faculties gained by its adventures in form. At the mental level this centre is termed the mental permanent atom, and it serves as a focus whence life unfolds throughout the general evolutionary scheme. Through this permanent atom, fohatic energy plays, and thrusting down, develops a further centre at the lower intellectual level which can act as a focus to sustain and organize form in the physical world. From thence the transitory forms of mineral, plant and animal life are built, and the vitality of the plant and the instinctual life of the animal is sustained. It is the one force, the fohatic energy that is the controller and organizer of all sentiency and consciousness and of those influences that serve to awaken such sentiency. Hence Fohat is termed at this level the organizer of forms and the life producer.

When animal awareness and an organism of sufficient

complexity and sensitivity has been developed it can be specially stimulated for the use of a self-conscious entity. Such an entity is a human being. He is linked to a developed animal organism vitalized from a centre at the denser thought level and thereafter uses the whole animal mechanism for contact with the substantial worlds. This mechanism is the persona, the mask, worn in each life, the so-called kama-manasic sheath or body. It too expresses the triple aspects of spirit, but with considerable distortions due to its animal origin and hence its automatic polarization to the densest worlds. So, within the personality, will becomes manifested as desire, wisdom as feeling, while the divine activity becomes ensouled in ideation—the ability to construct forms out of thought material and to give them objective existence and appearance.

The denser mental world (kama-manasic) is the field whence is energized the biological activities of living forms. From it are generated those laws which affect life and form and which maintain harmony between the form or structure and its environment—a harmony which is technically termed the 'function' of the organism. In so far as man uses the body of an animal he is also subject to the laws which rule all living organisms from cells to the higher animal types.

The lowest region of all is that of matter and force, and concerns the 'wondrous laws of Matter'. *(The Secret Doctrine,* III, 37.) Here Fohat manifests as force. Most of the ordinary laws known to physical science deal with this level and they can be grouped under one or other of the gunas—acting at this level—if the mode of motion which predominates in the expression of any particular law can be determined.

The following diagram sums up the operation of universal law in relation to forms and to consciousness at the three levels of spirit, mind and matter Expressing itself characteristically as consciousness at the spiritual level; as what might be termed feeling-thinking, and life-force in the dual mind; and as force-matter at the purely physical level; the universal force then develops consciousness into self-consciousness, feeling-thinking into a vivid awareness of living, while at the lower levels it maintains life and form, and force and matter in harmonious interaction.

Some of the laws of ordinary science are shown grouped under one or another of the gunas. Any grouping of this kind must be, however, of a very tentative nature at present.

Table IV
THE ONE FORCE AS THE ORGANIZER OF FORMS

Region	Characteristic Expression	Faculty Developed			Energizing Power
Spiritual	Consciousness	Self-consciousness			Fohat as the Light of Consciousness
		(1) *First Aspect* Will	(2) *Second Aspect* Wisdom	(3) *Third Aspect* Activity	
Intellectual (a)	Feeling-Thinking	Awareness of Life			Fohat as "The Fiery Steed"
		(1) *First Aspect* Desire	(2) *Second Aspect* Feeling	(3) *Third Aspect* Ideation	
Intellectual (b)	Life-Form	Organization of Form			Fohat as Life-Producer (Manifests Life)
		(1) *First Aspect* Wholeness of organism	(2) *Second Aspect* Balance of Organ and Function	(3) *Third Aspect* Progressive expansion through form	
Physical	Force-Matter	Manifestation of Force			Fohat as Force
		(1) *First Aspect* Law of Limitation, Sacrifice	(2) *Second Aspect* Law of Harmony	(3) *Third Aspect* Law of Multiplicity	
		Laws of Statics, Gravitation	Laws of Harmonics	Laws of Dynamics	
		Rotary motion	Vibratory motion	Translatory motion	
		Tamas-guna	*Sattva-guna*	*Rajas-guna*	

Although, for the purposes of a diagram, the levels of consciousness are arranged in a vertical sequence, it is apparent that they are more truly the expression of a within-without relationship, or a 'hierarchy of levels'. A complex being such as man, with an unfolded centre at the spiritual level of self-awareness, acts from within, through the centre of personal incarnation in mind and an organized bodily form, out into the physical environment upon which he exerts himself by the use of his organs of action and reception. These latter are instruments for maintaining the force-matter interaction.

An animal consciousness with its dawning sense of individual focus at the mind-feeling level (later to become the personality-centred human instrument with its fierce sense of I-ness), expresses itself likewise through its bodily form, thus affecting the surrounding environmental field. Even the simpler plant organism has an inner focus of organized life from which the outer form is controlled. Successful and harmonious living consists in the right ratio and balance of this within-ness and the organized form which it has created for its use, together with a balanced relationship between the organism and its environmental field. The triple rhythm of the cycle of necessity, outlined in the introductory principles, causes every organism to pass through cycles of changes, so that the outer form as created for the use of the life manifests for a time and then passes away. As Tagore has expressed it 'Life which is an incessant explosion of freedom finds its metre in a continual falling back in death'. All incarnation in form is subject to this cyclic process in which life is succeeded by death, only to renew itself again in a wider expression of life. Rightly understood, the release of life from form is a process of regeneration and transmutation that is infinitely repeated, and can be experienced in every death of form when life transcends its outworn envelope. The casting off of an outworn form, is, or can be, a small foretaste of the true turiya state at the end of a major cycle. For a moment the weight of form is lifted from the life; activities subside.

Sattva alone remains and the peace of the spiritual level dominates. Then the next cycle again demands limitation and activity, and so the process continues with a new plunge into the cycle of necessity.

MATERIAL AND SPIRITUAL TRANSMUTATION — THE TRANSCENDENCE OF LIMITATION

All incarnation in form is a continuous process of transmutation in which there is progressive and dynamic change. Creative living is concerned with the assimilation of the essence of experience in limited forms, experience developed through the interaction between the 'vehicles' employed and their surrounding environments. Thus to live creatively is the supreme expression of the fourth phase, wherein the higher octave of sattva brings all the three qualities into a whole and harmonious relationship.

At the level of biological life an animal body lives by a process of intake and rejection and by the absorption of food which provides its raw materials. By a process of assimilation this is transmuted into the living substance of which the body is composed, and waste products and unwanted material are thrown out as dross. The process is dynamic and selective and involves a continual breaking down of material forms and the extraction of their essences. Controlling and harmonizing this process there is, in a healthy organism, an impulse from an inner subtle level acting outwards into the denser medium.

Thus any living organism is maintained by a threefold relationship. There is first the living entity using the organism, a part of whose being is at a subtler level than the dense physical; this employs a body at the physical level closely linked with, and controlled by, that living entity, and the body is in its turn immersed in a surrounding environment or medium within which it grows and from which it derives nourishment. The organism achieves success in its living as it succeeds in thrusting outwards to make contact with the environment and in transmuting the material supplied by the environment field into energies for maintaining its form and carrying out its own functions. Such transmutation involves both the experience of a two-way interaction of the body and its environment at the level of form, and a distillation or extraction of the essence of that experience, at an interior level, at which the entity using the organized form has its being. This distillation of experience stimulates a further thrust of activity, of power, from the living entity out into its form and hence to the environment. It expresses a will to live; it is engaged in maintaining itself in a body and in holding the balance of that body with its environment in a correct selective and rejective ratio. An understanding of this dual interaction is the key to the mystery of transmutation. The body gives and receives in relation to its environment; it also gives and receives in relation to its inner focus, its centre of life. The importance of the play from the inner centre outwards increases with the complexity of the form. If more and more of the powers of the energizing life of the organism can be brought to bear on the environment through its form, then the possibility of creative activity within the environment arises.

Simple plant and animal organisms maintaining themselves at the physical level somewhat at the expense of their environment, represent perhaps the most rudimentary expressions of this activity, which were it not for the competitive claims of other forms, might choke the world with one type of organism. This tendency continues throughout the kingdoms of nature in even greater complexity until the whole intricate web of reactions is built up by which a human being maintains himself from an organized centre in the intellectual realm and from there acts outwards to dominate his physical environment.

There is yet another and far more significant use of this two-way process which is a further expression of the law of harmony or equilibrium, and which can permit man to express his true spiritual will within all the regions of experience and so live as a conscious spiritual being. Every human being is, as a centre of Logoic life, a spiritual focus of the one atmic power of will. Each has a centre of awareness in the spiritual world, more or less awakened and clearly defined, and at some times more active than at others. This corresponds in its relationship to the outer man to the inner focus of being of the living organism. As the inner centre is the sustainer and energizer of the dense body, so is the spiritual will deep within the human being the source and energizer of his whole nature. Without its existence he would cease to be an entity at all. From the deep level of spirit the will supplies spiritual energy for the creation of forms. It thus isolates a portion of the middle intellectual field of experience, to serve as a body or vehicle to be its expression at the level of mind. This is the mind-emotion body or kama-manasic sheath of *The Secret Doctrine*. Like the physical vehicle in its physical environment, it is made of the material of the field in which it operates. And as the physical body of a living organism has the peculiar vital quality we associate with being alive, so the mento-emotional body has a living quality which distinguishes it from pure mind-substance. It behaves as a living entity, as does the physical body, and is endowed with desire and a vitality of its own. We term it a 'personal elemental'. There is an interplay of assimilation, of intake and output, between this 'elemental' and its environment, two-way like the corresponding biological relationship of the living organism. This interplay is the complex process of thinking-feeling; sensing, receiving and absorbing the thought forms of others, throwing off and energizing consciously created, or more unusually, automatic superfluous thought forms of our own. Furthermore, just as the inner life-focus automatically controls the breath or outflowing life, sustains the physical organism and at the same time receives the distilled essence of physical life and vitality, so the spiritual focus or centre interacts in a two-way manner with the mento-emotional form.

For most of humanity the inner centre of spiritual power is subjective and latent. Spiritual transmutation and the re-establishment of the higher sattva, of wholeness or harmony, begins when we realize its existence and its power. A real change in the focus of consciousness is involved. So often, we human beings engage in much mental and emotional activity, wishing and desiring to be other than we are, thinking of plans, schemes and systems for regenerating the world and ourselves; but we fail to make the essential shift of focus from the level of thought and feeling to that of the will. Without this all our efforts are useless and we are as helpless as is a user of electricity when the power is cut off.

The centre of life has to be consciously shifted inwards to the spiritual world, and the power of will realized there in the buddhic light. Thence it may be directed outwards gently into the personality in the intellectual world. Thus employed it can act as a shaft of living lightning to break up old habitual patterns at the mento-emotional level, release the life locked up within them and free it for self-directed activity. Thus is the great alchemy of transmutation accomplished. Thereafter the personality shines only with the light and life of the spirit instead of by the reflected ray of its own rajasic activity. It then can become truly self luminous and sattvic. Further consideration of this, the great act of transmutation to be achieved by humanity, is the theme of the next chapter.

CHAPTER 10

MAN, THE MEASURE OF ALL THINGS
By E. L. Gardner

'The Universe is worked and *guided* from *within outwards*. As above so it is below, as in heaven so on earth. "Man," the microcosm and miniature copy of the macrocosm, is the living witness to this Universal Law.'
The Secret Doctrine, I, 317.

'There is one Eternal Law. . . . It is owing to this Law of spiritual development superseding the physical and purely intellectual, that mankind will become freed from its false Gods, and find itself finally—Self-redeemed.'
The Secret Doctrine, III, 418.

NATURAL LAW

THE FORCES OF NATURE are said to work according to law because their behaviour seems to be invariably consistent, as indeed we find them to be within the limits of our everyday experience. This consistent behaviour has resulted in the framing of laws such as those relating to gravity, inertia, thermodynamics and the like. Accuracy in their observation, however, is now recognized as being wholly dependent on the nature of the observing consciousness. The law of gravity, for example, would be defined very differently from the viewpoint of an insect or a fish, compared with the human view. So also if the earth became heated to the temperature of the sun or chilled to that of, say, liquid oxygen, our reading of natural law, assuming we had bodies corresponding to the environment (vide *The Secret Doctrine*, III, 224), would need much amplification and revision.

Our interpretation of natural law in its widest application depends, in short, on ourselves and our modes of response. Conquering nature, by obedience to natural law, implies the adjustment of man to his environment in a harmonious relationship that must include both the physical and metaphysical factors of human experience. And the human perception of natural law will always be relative and partial, depending on the evolution of the human consciousness and its relation to its surroundings. But in *The Secret Doctrine* we find the statement—

> 'There is one Eternal Law in Nature, one that always tends to adjust contraries, and to produce final harmony.'
>
> *The Secret Doctrine*, III, 418.

The pressure of this 'eternal law' is elsewhere, and more usually, called karma,[1] and, for most of us, such an inclusive statement as the above can be accepted intuitively only; present human consciousness is too limited to survey the vast cycles of time that might demonstrate its truth. Moreover it has a certain ethical connotation, a purposive quality in relation to consciousness, that is not evident in the laws that are observable by the physical senses. Yet, if it be granted that manifestation follows *design*, then the existence of such an 'external law' is soundly logical.

Our concept of a law that is universal will widen, of course, as consciousness is able to observe objectively more and more of its effects in the field of emotion and thought. In the younger kingdoms the automatic operation of given laws is to be seen in the instinctual behaviour of birds and animals and also, extensively, in the psychological reactions of human crowds under a strong

---

[1] Karma: usually stated as 'Man reaps that which he sows'. Is related to the physical law that action and reaction are equal and opposite, that cause is followed by effect, etc.

stimulus, and in psychotherapy generally in the practice of suggestion and analysis. Certain levels of these subtler forces of natural law are being studied now under laboratory conditions by enquiring medical and physical scientists. Thus our concepts of the field of law widen as human consciousness advances — though they are still conditioned by the limitations of the perceiving intelligence.

In the critical period of the world's history, through which we are now passing, the place of man needs study and interpretation. It is his task and his alone to observe, understand and apply the laws of nature to their ultimate uses, not only within the everyday physical world but also at the subtler levels of emotion and thought. When it was declared that 'The Breath needs a Mind to embrace the Universe' *(Stanzas of Dzyan)*, the making of man was begun — and the tremendous task thus set for humanity explains the otherwise startling dictum that all life must pass through a human kingdom to acquire a consciousness that may become immortal. Such a consciousness is the dharma of man as he learns to obey — and thereby transcend — the universal law of karma.

THE HUMAN KINGDOM

The two to three thousand millions of years that are now reckoned to be the age of our world[2] are but a part of the cycles of time that occult research claims for the making of man — and perfect man is definitely asserted to be the goal and consummation of manifestations within our solar system.

> 'The phenomenal world receives its culmination and the reflex of all in MAN.'
>
> (III, 48.)

> 'Man is the microcosm of the universe.'
>
> (I, 230)

---

[2] The age of the Earth is now believed to be about 4.5 thousand million years. The Universe is probably several times that age.

> 'Man ... the "storehouse" ... of *all the seeds of life* ...'
>
> (III, 291.)

> '... from the beginning of the Round, all in Nature tends to become Man. All the impulses of the dual, centripetal and centrifugal, Force are directed towards one point — MAN.'
>
> (III, 177.)

> '... Spirit and Matter, the union will produce that terrestrial symbol of the "Heavenly Man" — PERFECT MAN.'
>
> (I, 292.)

Such are the statements. If man constitutes such a critical turning point in cosmic history, we must take a far longer view of his past than is customary in any work of modern science dealing with human origins.

ORIGINS AND RECAPITULATION

A well established view in biological science is that the history of the growth of the individual, from the earliest gestatory period to maturity, is a shortened recapitulation of the whole long-time history of the species. Relating this to man the implications are searching, for this leading principle of growth must apply to all the planes of being. The present-day human physical body begins as a minute single cell and develops through multi-cellular forms as an embryo, a foetus, and on to the new-born child; on its way it passes through the same phases as the younger kingdoms of nature, emerging at the end of a brief nine months of gestation as a human being. The theory, based on occult research, that the forms of all the kingdoms have been first outlined by man, in the subtler planes and in the physical, has much support from these progressive sequences of gestatory growth, the theory being that at the completion of certain cycles the form then reached by the human pioneer was discarded and left for a younger following life to occupy. (The pioneer humanity who led the earlier developments were of a preceding manifestation, not our own; they were of the successes of past cycles who chose the path of service and filled the office of leaders for the following life waves.)

In the making of all forms, the inner subtler planes of the mental and astral worlds are the first to be used and they are readily responsive and obedient to inner direction. Just as form is *rapidly* built now during gestation from a single cell onwards to the finished body of man today, so were the earlier subtle forms *slowly* developed and periodically discarded through the millions of years during which, and even before, our present planet, the earth, was assembled. And, correspondingly, every time a human being seeks a new incarnation he descends similarly from higher planes (wherever devachan was spent) down through mental and astral levels, gathering sheaths from both for his new astro-mental body, and on to the physical for the protective 'coat of skin'. The course followed each time during the individual's descent to another life on earth, nowadays comparatively swiftly from highest to lowest, is a shortened recapitulation of the long ancestral descent of the species.

In the subtler and physical bodies of a new-born infant, there is thus concentrated the work of literally millions of years of labour on the part of the devic form-builders conjoined with the 'will-to-live' of the monads of our hierarchy. During these vast cycles of development, the successful progress made has resulted from innumerable repetitions: and the marvellous skill in craftsmanship of the rupa-devas of today enables them to build in a few months forms that have taken untold ages to model in detail.

The Monad's powerful will to live, *consciously to live,* is the answer to many a query that is posed today, such as— Why all this immense manifestation of activity, this seemingly endless struggle and strife with burdensome effort and brief pleasures, to achieve at length a mere release from a cycle of birth and death? The following emphasizes this answer.

> 'For the Host that incarnated in a portion of humanity ... preferred freewill to passive slavery, intellectual self-

> conscious pain . . . to inane imbecile instinctual beatitude.'
>
> (III, 419.)

Therein is the difference marked between a life that is a mere living under the constraint and bondage of nature's laws and the freedom of self-conscious, self-directed, life. And to attain that freedom, a Self-generated effort is imperative: hence the occurrence of crises in the development of true human consciousness, crises that may be turning points to great success or woeful failure.

TWO MAJOR CRISES

In the approach to a new incarnation, during the rapid recapitulation of the involuntary past, a major critical phase is the union of the monadic life 'descending' with the 'ascending' infant body. The crisis is touched at the quickening in the fourth month of gestation and is passed with a successful birth. In the third root-race on earth, this union of human life with a dense form took numberless experiments of trial and error and skilful adjustments. The ease and rapidity with which this is accomplished today is due to the skill of the rupa-devas in the grafting, to borrow a botanical term, of a lofty 'scion' to a sturdy 'stock', in uniting 'highest spirit' to 'lowest matter'.

The first major crisis was thus successfully passed in the third root-race and, with this achievement of union between highest and lowest, much else followed: the coming of the Lords of the Flame to take over the world's government; the stimulation and development of the mental principle in man; the division of the sexes; the beginnings of family and tribal responsibilities; the use and control of fire, etc., etc. This cosmic success in our third root-race can be taken as ushering in the final stages of the involuntary arc of descent, at the end of which spirit is immersed in matter and form. Doubtless, correspondingly, it represents the climax to preceding stages during which — in the third Chain and the third round of our fourth Chain — preparations and rehearsals on a vast scale were made for this dramatic finale in our third root-race.

On the evolutionary arc of ascent, the corresponding climax will occur in the fifth round and constitutes the second major crisis. This has been often allegorized and appears in religious references as a day of judgment. The point that affects us here and now is that—again correspondingly—a crisis is reached in this the fifth root-race period on earth by way, again, of preparation and rehearsal for the prospective finale of the fifth round. This minor crisis, as it may be called, applies particularly to the individual—its appeal is to the few who choose among the many who may hear. These crises, both hierarchical and individual, are symbolized clearly in a triangular diagram.

THE TRIANGLE DIAGRAM

The path taken by the Monads of our own hierarchy, 'journeying through the worlds' is represented—in an all-over inclusive view—on the three sides of a triangle. The descending line on the left, from apex to base, indicates the outgoing, or descending, path of involution. The base-line from left to right tells of a state of balance wherein spirit is at grips with matter on practically equal terms. The returning ascending line from base to apex, on the right, indicates the evolutionary path whereon spirit becomes ascendant—with the fruits of manifestation in the terms of consciousness.

Now let us trace, in more detail, the path taken by the individual who, in common with the hierarchy as a whole, is repeatedly recapitulating the ancestral course. The descending left-hand line of the triangle represents his entry into the worlds of form where mental and astral sheaths are gathered around the permanent atoms as nuclei. The physical etheric field is entered next and the first critical point is reached when the dense physical body is adopted. This is represented as turning the acute angle on the left hand and successfully incarnating. The bodily elementals' craving for food and shelter and sensual gratification provide the incentive that persuades, induces, 'tempts' the monadic ray, the ego, to identify itself with its bodies. Particularly useful, though intensely limiting, is the dense body because it provides a firm and stable anchorage and a fortress of protection while the subter planes are being consciously explored, used and controlled.

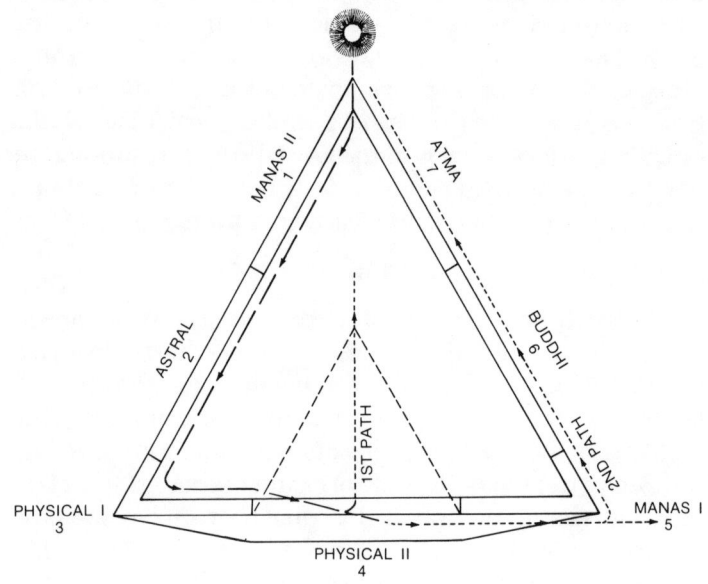

Diagram II

Throughout the critical period of balance between spirit and matter, represented in the diagram by the baseline of the triangle, the physical body is repeatedly used in successive incarnations. And the constant claims of the body for supplies of food compel its owner's attention and strengthen his link with it. The eastern term of 'food-sheath' for the physical body is very apt. Prana, the throbbing life-principle of all the planes, is amply sufficient for the subtler bodies but, in physical terms, prana is directly potent only in the flow and ebb of breathing. For the new exterior physical form, new not only in this the fourth Chain but new with regard to the whole Terrene Scheme, something much more is demanded; elaborate processes of mastication and digestion are necessary. As the body becomes more sensitized and refined, nourishment advances from the omnivorous to the produce of the plant kingdom alone. Eventualy, in the sixth root-race, nutriment is derived from fruit, pure fruit only. The sequence appears to be, omnivorous, vegetarian, fruitarian and, finally, the essence of the latter probably in some subtle aromatic prana that may eventually be derived from the grape or an allied fruit. There have been many speculative, and practical, experiments along these lines, prompted by the belief in an elixir of life, but, obviously, the physical mechanism must be commensurately prepared.

The halfway point along the base-line of our triangle, or thereabouts, represents the place where the majority of our hierarchy stands at present, and its future path is shown as continuing to the acute angle at the extreme right. This turning point, the second critical phase, will be reached, as has been stated, in the fifth round — still a very long way ahead. The task to be accomplished on or before then is the attainment of self-consciousness at the higher mental level and the consequent ability to function in waking consciousness in the causal body.

The pressure of nature's laws, and that aspect of universal law known as karma, operating on all the planes will tend continually to guide and shepherd human progress, though at a very, very slow pace. This gradual and slight incline 'upwards' is said to be the average mode of advance; it has been likened to the climbing of a mountain by winding around on a slowly rising path. There is, however, another path which, adopting the same illustration, may be said to mount perpendicularly to the summit. The goal of the Chain is the same for both paths, namely, the enlightened consciousness of the Asekha Adept, corresponding to the Fifth Initiation.

THE TWO PATHS

The awakening of the high principle of will in man makes it possible for choice to be exercised. The shorter, and harder, way to the summit may be chosen, if desired. The veil of emotional, personal desire, that has been so long a veritable ring-pass-not, can be rent — by the will.

Human consciousness, when crowned by its highest principle, the will, can transcend the limitations imposed by nature's laws — the laws that govern automatic activity, and so can rise above the traditions and conventions that have been built up around the personality and which tend so thoroughly to cloud and prejudice the mind. Consciousness, endowed with will, can touch and release a source of solar and universal power hitherto scarcely known. This power is of the First Aspect and enters at the gateway of atma. Though at this level but a thread may be touched by consciousness, it is ample to dissolve all that is outgrown and outmoded and, in a measure, begin to clear the mind of its clouds. This is the aspect of will that acts as a destroyer. As obstacles and fetters are removed and loosened that which karmic law can evoke but slowly is hastened. This is the promise to be fulfilled with the advent of true free will.

There is no contravention of natural law here, but the introduction of a new force that is greater and subtler than the law's gentle and persistent pressure. Man alone, on earth, can exercise this power. With the awakening of the true will in man, the choice between two paths lies ahead!

These two paths are shown diagrammatically, by arrows, on the triangle. The one, indicated as travelling on to the right-hand end of the base-line, is said to be taken by the large majority of our humanity. The first and shorter path is shown as striking directly upward, springing from the mid-physical level. This is the choice made, through an act of will, by a minority, by 'the few'. This direct path is led by the most advanced of our hierarchy, of whom some are now in high office in the Inner Government, together with the great Adepts known as Masters of the Wisdom.

THE CHOICE

The paths are usually described as —

No. 1   The Way of the Heart, and

No. 2   The Way of the Head.

Taken literally, these words, head and heart, are somewhat misleading because the 'heart' of the four-fold personality, the nuclei of the soul (permanent atoms), is closely related to the 'hollow of the brain' wherein are the pituitary and pineal bodies — and both therefore are in the head. The doctrine of the eye (more apt than the 'head') and the doctrine of the soul, or heart, are also descriptive terms used for the two paths, and apply respectively to an exoteric and an esoteric approach.

The following verses from the *Book of the Golden Precepts* (*vide* the *Voice of the Silence*), define the paths —

'The Dharma of the Eye is the embodiment of the external';
'The Dharma of the Heart is the embodiment of divine wisdom';

'The Doctrine of the Eye is for the crowd';
'The Doctrine of the Heart for the elect'.
'Follow the wheel of life; follow the wheel of duty to race and kin. Exhaust the law of karmic retribution'.

'If sun thou canst not be, then be the humble planet'.
'If thou art debarred from flaming like the noon-day sun . . . then choose a humbler course.'
'Point out the way, however dimly, as does the evening star to those who tread their path in darkness.'

'If the doctrine of the heart is too high-winged for thee . . . remain content with the eye doctrine of the Law.'
'If the secret path is unattainable this day, it is within thy reach tomorrow.' (Tomorrow = the next incarnation.)

A third path, if it can be so called, is tentatively shown on the diagram as branching off from the angle on the right. This path of 'failure' relates only to those unable to reach and pass the test of the Fifth Round and who, from then on, await another field of manifestation.

THE WILL

'The Tamas of the higher self is the Spiritual Will that knows itself and can never be shaken. The Will is the strongest and quietest principle in manifestation in Man. He who has reached that in himself is seated as on a rock—for he knows, within himself, that the divine purpose is never thwarted.'

SUMMING UP

Power—Fohat—used in the manifestation of the solar system, modified by the Will of the Lord of the Sun, becomes His 'one eternal law' within the solar ring-pass-not. His law makes for harmony in relation to life and form, harmony of interaction everywhere.

The power delegated to the Rulers of His planets, including that within the particular ring-pass-not of our world, is used and differentiated by the devic lords of the elements and operates, for us, as the laws of nature.

Of all life in nature's kingdoms, man alone can consciously live in the two spheres, the terrene and the solar, but not till man's highest principle, the will, is in some degree awakened, is he free to use the solar power which endows him with freedom of choice.

Freedom to choose, detached in a measure from the terrene personal atmosphere, can be normally exercised after the completion of human *in*volution, that is, after the midway point is passed in the foruth root-race on earth.

The use of the will to choose develops slowly, but during the fifth root-race an important opportunity occurs because the minor cycle of a 'fifth' corresponds in its purpose and qualities with the much later fifth round, on the cycle of return, when the causal body of man, of buddhi-manas, is to be self-consciously used.

The choice that is thus open lies between two paths of further human progress, known as the Way of the Eye and the Way of the Heart, the exoteric and the esoteric respectively. Both lead to the same goal but the one, the exoteric, is the long and winding road 'round the mountain', subject constantly to the pressure of terrene, or natural, laws. The other, the esoteric, is direct, 'uphill all the way' to the summit and demands for its ascent the use of the solar power under the creative drive of the human will.

The goal may be attained under the normal pressure of natural law and the subtle inexorable persuasions of karma or, with a definite expression of will to sustain continued effort, nature's laws, that govern the whole personality, may be transcended and the swifter path chosen.

This latter, the path of direct ascent, is a way of strenuous concentration and disciplines. It is the path of discipleship and initiations.

CHAPTER 11

BEYOND LAW

by NEVILLE REED

> 'The Occultist . . . enters . . . on to the path of individual accomplishment instead of mere obedience to the genii which rule our earth. This raising of himself into an individual power does in reality identify him with the nobler forces of life, and makes him one with them. For they stand beyond the powers of this earth and the laws of this universe. Here lies man's only hope of success in the great effort; to leap right away from his present standpoint to his next, and at once become an intrinsic part of the divine power as he has been an intrinsic part of the intellectual power, of the great nature to which he belongs.'
>
> *Light on the Path*, 61, 1909 Editon.

THE LAWS OF EVOLUTION, karma and reincarnation are expressions of universal law. These laws provide a convincing explanation of life, but perhaps a misleading one. In the welter of law-abiding phenomena one is apt to overlook the exception to the rule, namely life itself. Life leaves us alone to progress towards life. It does not concern itself with individuals; if it did we should be quite different from what we are, we should be merely a collection of perfect will-less robots. We are not robots, we are men, Gods in exile, Gods in chains, chains of our own making. A clue to man's great opportunity is contained in the myth of Prometheus. Man has forgotten his divine origin through long enslavement in transient things. In one sense, we do not evolve, we remember.

KARMA AND GRACE

The law of karma is certainly true, but it is only a half truth. It does not give us the whole story; we are apt to regard it as all binding, but this may not be so in every case. Individual unaided effort can take us so far, but no farther. The kingdom of heaven cannot be taken by storm. There comes a time when the initiative most come from above, when another factor 'outside the law' takes control. The old conception of grace that follows repentance and remission of sins is more than a myth. There is room for the operation of both karma and grace in the world.

To assist us in breaking the hold that the conception of karma may have over us, it is helpful to consider the analogy of the dream state of consciousness. In a dream we see objects and meet people under conditions that present some degree of law and order *because they are held in the embracing mind of the dreamer*. These apparently rigid laws of the dream turn out, when we awake, to be only self-imposed limitations.

THE ILLUSION OF MATTER

When considering the relationships between thought, emotion and the physical state, such as the action of mind on matter, we are apt to regard matter as something that is fundamentally different from, and can exist apart from, mind. It will be shown that this is a false, if excusable, point of view. Those who look for dramatic evidence, supporting the action of mind over matter, such as cases of materializations, are liable to miss the wood for the trees. They fail to realize that every time we raise an arm it is a case of mind acting on matter.

The implications of being able to affect matter (our physical bodies) by our mind are tremendous. If a cricket bat is able to affect the path of a ball it is only possible because they are both solid objects. Now we know that if we want to raise an arm we can do so. Therefore our mind can affect matter. It follows that *mind and matter must fundamentally be of the same nature*, otherwise they cannot interact. Either mind is a form of matter, or matter is a construction of mind. We have only indirect knowledge of the outside physical world, but direct knowledge of our own thoughts. Therefore the second alternative must be the true state of affairs, namely matter is a construction of mind.

The conception of matter is an attempt to explain why different individuals have similar experiences; nevertheless, the existence of matter is only an assumption, and although it is useful for practical purposes it introduces more difficulties than it solves. There is an alternative explanation to account for the similarity of individual experiences that does not require the assumption of matter as an independent entity: namely that we all share and participate in an embracing consciousness. The world is real, but it is a mental reality, not a material reality. So long as we believe in the independent existence of matter we remain materialists. Let us take the plunge and rid ourselves once and for all of the Maya of matter. What we used to regard as matter is in reality the individual mind becoming aware of a univeral thought. Dreams, imaginations, etc., are creations of the individual mind, our experience of Trafalgar Square is a construction derived from universal mind.

One of the functions of Theosophy is to combat the wave of gross materialism that is spreading over the world. This must be accomplished by revealing the mental nature of matter, not by a retreat from matter. Where religionists are mistaken is that they fail to realize that in its basic nature matter is just as divine as spirit. We are only aware of ourselves and the outside physical world because we are thought-held in an embracing consciousness; every minute of the day we have visible and living proof of the oneness of life. Behind the outward appearance there is an inner reality which mankind must learn to understand and to *love*; this is the hidden message that life is silently confronting us with all the time.

THE ILLUSION OF LAW

The laws of evolution, karma and reincarnation are misleading in a similar manner, and for precisely the same reason that Newton's laws of physics are misleading when compared with the theory of relativity. They ignore a fundamental factor, namely the observer. It may be difficult for us to realize the fact, but matter cannot exist on its own. *Nothing is known apart from the knower.* The universal laws are known only because some mind is aware of them. The object of observation and the observer cannot be separated, they merge into a unity, a state of awareness. The general, but mistaken, point of view is that man is placed in a world like that which he perceives, and then proceeds to study it; but in reality it is just a model which he himself has placed on the scene. To get a true picture nothing less than a complete somersault of outlook is required.

When we observe the heavenly bodies we become aware that they all participate in one universal motion — a revolution round the polar axis. But we know that this one great fact which is true of them all has in reality nothing to do with them, it is the result of a condition of the oberver.

The above is an example of a profound truth. The more universal a law appears to be, the more certain we are that it is a property of the observer. It will be seen that all this leads straight back to the familiar 'man, know thyself.'

THE ILLUSION OF SELF

If we try to understand ourselves we find that the majority of our thoughts, feelings and actions are motivated by a craving to become. We strive after one particular goal or try to avoid another, always struggling to become something different from what we are. This striving to become shows itself in a craving for pleasure, a craving for power, and a craving for immortality. It is the last that most concerns the subject under discussion. We satisfy our craving for immortality at the price of not understanding the present. There is a risk of getting out of phase with life. It is well to remember that the purpose of Theosophy is not so much to spiritualize, as to humanize us. There is a general but mistaken belief that progress is made by a movement away from matter towards spirit, whilst the correct conception is one of movement away from personal selfishness towards impersonal unselfishness.

It must follow from the coequal existence of matter and spirit and their interaction to produce a manifested universe, as outlined in the second chapter, that in the marriage of spirit and matter, matter is an equal partner in the union, and should not be dominated by spirit. There could be no greater travesty of the purpose of life than to regard it as one of rapid retreat to Nirvana. There has been much emphasis on the disastrous consequences of a life spent in pursuit of pleasurable sensations; it is well to remember that the other extreme may be even more dangerous. Many lives (of the ego) can be lost by killing out the desire for life, long periods of time then being spent in a state of suspended consciousness.

The concept of evolution may lead us to believe that our salvation is to be found either by retreat into higher states of consciousness beyond the gross physical world, or at some remote period in the future. Nothing could be further from the truth! Only here and now have we the opportunity to fathom life's greatest mystery, to take the essential step.

Throughout life we gradually learn that first appearances are deceptive. The common sense viewpoint is continually being overthrown by reflective reason and experience. The earth seems flat and stationary but we soon learn how misleading are such apparently obvious conclusions. It is difficult to realize the world of solid matter outside us as a construction of mind. There is yet another violation to our common sense. The one thing above everything else that we are sure of is the sense that I am I. However other people may persuade us, they will never shake this rock-foundation of our nature. In our grasp of the sense of I-ness we are caught like a monkey in a trap. The trap consists of a jar containing a nut and tied to a tree. The monkey can get his empty hand through the neck of the jar to grasp the nut, but cannot pull it out unless he first lets go of the prize. The monkey will not let go of the nut and so cannot attain his freedom. Of a like nature is the problem of the I in man. Will we ever drop the greatest of the illusions, and let go of our sense that I am I? Or, the same thing in another form, expand the I to include the you? Hints of the quest are revealed in the myths of the Holy Grail, the Flying Dutchman, and many other legends with a sacrificial theme. The key is to be found in the right understanding of the much abused word *love*.

The value of the study of universal law lies not in satisfying our cravings by bringing us tidings of comfort and joy, but in its ability to awaken in us a realization of our true nature.

Granted that we are all aware, to a greater or lesser extent, of some underlying law in the uiniverse; in other words that it is a cosmos and not a chaos; what is the significance of such a law? Surely it is that the individual and the outside world, the 'I' and the 'not I', are artificial distinctions; an illusion due to the limitation of the individual.

CHAPTER 12

IMMEDIATE APPLICATIONS

by E. Lester Smith

IN THE PREVIOUS CHAPTERS we have deliberately tried to describe the laws of our Universe in the broadest possible terms. Our survey has ranged through all the levels of manifestation, and through long aeons of time. Accordingly it is bound to seem abstract and perhaps abstruse, and its apparent unconcern with immediate world problems may engender a sense of impatience. The purpose of this chapter is to suggest that these troublesome everyday realities can profitably be laid alongside the ultimate Realities, for more just appraisal.

It is proper for the Theosophist to have his head in the clouds — provided his feet are firmly planted on the earth, and love fills his heart. Only then can his hands be dedicated to effective service; service that is neither practical but uninspired, nor yet inspired but unpractical. Only such a man, spanning heaven and earth in his understanding, can with a sure grasp draw the triumphant future towards the troubled present.

There are plenty of practical men, working devotedly to solve the world's problems; but their imperfect appreciation of Universal Law stultifies their efforts, so that sometimes they only plunge the world more deeply into crisis. There are idealists, too; but they tend to over-emphasize one aspect of the Law, and sometimes seek to apply it out of season, with imperfect appreciation of hard facts. But the world's desperate need is for practical idealists — men with enthusiasm for hard work, with organizing ability and a deep understanding tolerance for human nature, linked with impersonal love for humanity, faith in the Plan, and comprehension of Universal Law.

We are all painfully aware that this is a time of crisis in the world. So also is it a time to resolve the recurrent crises, a time of glorious opportunity for service, as seldom before. But in order that our action may be wisely planned, it is well to pause awhile (though not too long) for reflection. Study of the fundamental basis of order in our universe should suggest how we may persuade mankind to live in a more orderly way. Also it tends to lift our vision from the immediate sordid problems of our troubled world, towards the deeper issues that lie behind. Thus our sense of urgency, though undiminished, may become tempered with serenity and faith.

There is a grave disadvantage to an intellectual analysis of fundamental truths; for the scientific method serves only to illuminate them with the cold light of reason; 'the mind is the slayer of the Real'. The intellect will inform us rightly only if it leads us on beyond itself. Philosophical thinking is sterile if it is not warmed by the glow of feeling, the appreciation of beauty and harmony. So it is good to *know* the Law; but this is not enough. One must learn to *love* it also.

'Law' is perhaps an unfortunate word to select; it calls to mind man-made laws with all their imperfections. But Universal Law is not cold, mechanical, and harsh. Nor is Univeral Law just a subject for study. One needs also to cultivate deep feelings towards it. Truly it *is* something to wonder at, to reverence — its awe-inspiring scope, its all-embracing inclusiveness, its fundamental simplicity, ramifying into a million inevitable complexities, and resolving back again into that one dominant aspect, harmony.

Here is material for ample aesthetic satisfaction as well as mental stimulus. Only then, when both thought and love have been brought to bear, can one hope to understand the Law—to achieve the synthesis that culminates in that flash of buddhic light, when nothing exists *but* the law, not even oneself.

It is not even enough to know the Law intellectually, to love it, and to understand it intuitively; finally one must practice it. The two remaining levels of our being must be attuned to the Law—will, and action.

When the atmic principle, the will, is allowed to take the reins, one *lives* the Law. One gains complete unhesitating trust in the Law; one lives by faith—not a mere passive faith, not resignation nor fatalism, but an active faith. One's affirmation becomes not simply 'God's will be done', but rather 'God's will be done *through me*'. With all this achieved, right *action* at the physical level follows inescapably, as verdure follows the spring.

Having thus become imbued with understanding of the serene orderliness of our Scheme, when viewed with insight and detachment, we may ask ourselves, how does our understanding differ from that of the intelligent non-Theosophist? What have we to offer him that might prove both acceptable and helpful? How can we help to dispel the prevailing mood of cynical despondency, and engender a renewal of faith in humanity and its ability to win through to a spiritually triumphant future?

It is not the purpose of this present Transaction to answer these questions, though we may attempt this in a later one—but rather to throw out the challenge they imply. There is material in plenty to inspire a spiritual revival, if we could find words impelling enough. To recapitulate just two themes: Acceptance of the anthropocentric view of our scheme expounded by Mr. Gardner would bestow a new dignity on the human race and show it a new purpose, casting out the ignoble ennervating doubt whether we were anything more than an unusual species of animal of no particular significance in the universe. A surely-based and most welcome sense of security would arise from acceptance of the whole concept of Law ordering *all* things and not just physical events.

Is it perhaps along these lines that the western world might focus its religious yearnings? Christianity, as interpreted by the churches, no longer commands wide spontaneous allegiance. The theme of the Law might provide new inspiration—a teaching akin to the Lord Buddha's. It should have a strong appeal for a race still enthralled by science but now realizing its urgent need for spiritual guidance.

> Such is the Law which moves to righteousness,
>   Which none at last can turn aside or stay.
> The heart of it is Love; the end of it
>   Is Peace and Consummation sweet. Obey!

PART III

MAN'S EXPANDING HORIZON

(This Purposeful Universe)

Edited by
C. R. Groves and Corona Trew

A transaction of the Science Group of the
Theosophical Research Centre, London

INTRODUCTION

THIS TRANSATION is the third of a series linked by the common thread of a spiritual view of the world's origin and progress. In *This Dynamic Universe* the world was seen as an expression of one life-force, termed 'Fohat' in the occult terminology of *The Secret Doctrine*, and the many workings of that one force were explained. The world as process was studied there. In *This Ordered Universe*, the concept that law reigns throughout every region of the universe was set out, and the nature of some of the great laws was considered. In the present volume it is the hope of the authors to show that there is both purpose and plan behind the evolutionary process, and to trace something of the workings of that purpose through the kingdoms of nature which find their consummation in man. With the concept of evolution as a unitary process expressing an unfolding spiritual purpose it is hoped to show that design and purposeful spiritual intent are the warp and woof of life. The stresses of man's life can only be resolved by the recognition that man and the universe are one and not two. Professor Sir S. Radhakrishnan, one time Vice-President of India and noted scholar and philosopher whose range of thought covered both Eastern and Western philosophic thought, saw the course of history as the expression of a 'truly domestic drama, the tension between the limited efforts of man and the sovereign purpose of the universe. Man cannot rest in a discord, he must seek for harmony and strive for adjustment'. It is a basic tenet of the theosophical vision of life that man and the universe are in essence the same. Man the microcosm reflects or expresses the vast cosmic process of the macrocosm, and so must ultimately subserve *his* purposes to *its* purpose.

It is another basic tenet that the universe is spiritual in origin, deriving from an interior spiritual state of being, that it unfolds into manifested existence and experience by evolving order and pattern from material substance, in order to express spiritual powers and potentialities. The universe is whole and basically harmonious in design and purpose. Man is an integral part of that universe and thus shares in the basic harmony, and human happiness can only be evoked as men resolve their individual tensions within the 'royal purpose' of the whole.

Evolution as the expression of a spiritual creative process is, then, the subject matter of this Transaction, and it is hoped to show that the problems of mankind become capable of resolution in the light of the larger view here presented. The modern world awaits the birth of its soul, and it is the view of the theosophical thinker that the soul is a spiritual entity conforming in its nature and structure with the highest purpose and spirit of the universe itself. This Transaction is published in the belief that a consciousness of a spiritual purpose is the greatest need of the world at the present day; and that this purpose can be realized both in the universe at large, and in the lives of all individuals within it.

CHAPTER 13

THE UNIVERSE AS AN ENTITY

ALL KNOWLEDGE of the material worlds gained by scientific methods is, from the very nature of science, an external view. The criteria of its validity lie in its reproducibility, its reasonableness and its communicability, and in the fact that it is built up by logical expansion from one part of such knowledge to another. To this may also be added its usefulness. When we look at the fabric of science we see that all we really *know* of its vast construction are the impressions and pointer readings gained by our senses and by the scientific instruments which, in effect, are extensions of these senses, together with the associated pattern of mental images which these evoke. The whole wonderful framework of scientific thought is then built up to explain the pointer readings we have obtained, and to predict new ones which we may perceive in the future.

There are, however, other fields of experience, *not* dependent upon external sensory contacts and their associated thought processes, which are accessible to the human consciousnes. These are the direct realization of meaning and value and the communication of mind with mind—the experience of the harmonies of ideas and ideals, direct, immediate and interiorly real and utterly satisfying. From science we learn that electrons, displaced from their usual position relative to one another by the absorption of specific frequencies of electro-magnetic radiation and the scattering of other frequencies, are responsible for our perceptions of the colour of the sky and the clouds. One may marvel at the complexity of the processes involved, but the experience of the wonder of a sunset is direct, immediate, whole and satisfying. It is also communicable in the sense that it may be a mutual experience shared by those who perceive it. We may likewise analyse the frequencies of musical notes into intricate patterns on an oscilloscope, but the aesthetic appreciation of a symphony or concerto is immediate and can unite groups of utterly diverse men and women into an experience of oneness. Deeper still lie the experiences of the innermost spirit in man, those ethical values and altruistic feelings, which are the expression of the human spirit at its most sublime. This range of values needs no apology or explanation for those who can perceive it; the only query which may remain concerns whence such values are derived, and whether they can sustain the human soul through the gateway of death. Is there any permanence and abiding existence for the human spirit? It is the answer given to this question that either gives to life its ultimate meaning and purpose, or else makes it a transient bubble thrown up in the sport of the forces of the cosmos. The real point at issue between the materialist and the religious thinker lies here, and it is just here that the theosophical vision of the universe as the expression of a mighty purpose can give enlightenment and hope. The seeing eye may perceive the purposeful universe with its ordered levels of function within which different patterns of energy are expressed until it reaches its visible climax in the human form. The value of this viewpoint lies in its recognition that the human spirit is identical with the essential life principle which informs the cosmos. This may become a matter of realization for the individual; a realization which is progressive and unfolds into ever deeper overtones of spiritual experience. The meaning and purpose of the universe lies in man and the human pilgrimage is designed to make him aware of his true destiny and role.

The very word *uni*-verse is an intuitive recognition by the human mind that what it is contemplating is regarded as a unity however diverse may be the objects of which it is composed. The vastness of this universe—its extent in both time and space—seems to increase as man invents instruments of ever increasing range and penetration. Radio-telescopes yield information from even greater distances than the largest optical telescopes.

What these distances mean cannot of course be realized clearly, but we may remind ourselves when reading of millions of light-years, that a 'light-year' is the distance travelled by light in a year at its speed of 186,000 miles per second. This implies that observable astronomical events are not actually occurring at the moment of observation, but that they occurred in the past at a time which may be anything up to two thousand million years ago.

There are two chief theories of cosmic origins, each of which has minor variants. These two theories have been much discussed and popularized, and will only be briefly referred to here. They have been called the 'evolutionary' theory, and the 'steady state' theory, and have been expounded at length respectively by Gamow and Hoyle. Modern theory also includes the possibility that there may be millions of planets whose conditions permit the existence of some forms of living organisms.

It should be emphasized that what is being discussed under the 'evolutionary' theory is the *present phase* of the universe. Gamow for instance speaks of a 'pre-galactic past' before the formation of the primeval atom and speculates on the possibility that before this the universe was collapsing from a state of infinite rarefaction until it reached the condition of infinite compression when it exploded again. This idea of a pulsating universe has been discussed by more than one astronomer. Thus, Dr. A. Sandage of Mt. Palomar Obseratory suggests that 'the universe may be pulsating like a gigantic heart-beat'. He believes that the recession of the galaxies is slowing down and that, if this be true, gravitation will ultimately overcome the expansion and the universe will begin to condense again so that several billion years from now the galaxies will collide in a fiery explosion. 'Then there will be an expanding universe once more'. The similarity between this and the 'out-breathing and in-breathing of Brahma' of Hinduism, and the 'numberless universes incessantly manifesting and disappearing' of *The Secret Doctine*, is obvious. If the 'steady state' theory be true then the universe is not necessarily 'running down'. According to Sir Harold Spencer Jones *(Science News* 32), the much discussed Second Law of Thermodynamics is not of universal application. There is no overall increase in cosmic disorder because entropy, which for this purpose can be equated with disorder, is 'carried away beyond the cosmic horizon by the outward movement of the galaxies. This means that the physical universe is not on the whole running down to a less orderly state. On this 'steady state' theory it is evident that the whole universe shows something akin to the homeostasis (self regulating mechanism) which is a characteristic of all living organisms. Thus we find Dr. J. Z. Young writing in *Doubt and Certainty in Science:*

'Might we then take as our general picture of the universe a system of continuity in which there are two elements — randomness and organization — disorder and order if you like — alternating with each other in such fashion as to maintain continuity? Is it possible that the data of the astronomers can also be interpreted as a system of balancing, of maintaining a steady state just as our living systems do, with the characteristic that it builds up systems of order and then returns them to disorder?'

The whole cosmos is thus compared with a living organism which demonstrates the principle of homeo stasis even at the cosmic level. Now at the biological level the living organism 'sucks order' from its environment (this phrase is quoted from *What is Life?* by Erwin Schrödinger) and living matter does not obey the Second Law of Thermodynamics which applies only to a closed system. It seems therefore likely that the cosmos, far from being a closed system, is constantly infused with new energy at the inorganic level which enables it to continue without loss of order. Indeed it is not impossible that both theories are true, that a new cycle begins with the explosion of the highly condensed 'nuclear fluid' of the evolutionary theory, but that during its long life the steady state theory holds, until a cosmic contraction begins and finally ends the cycle.

It has already been suggested above that the universe, in spite of its multiplicity, is in fact an integral whole whose diversity is rooted in, and owes its very existence to, an underlying unity. That this is so in respect of material constitution and of physical 'laws' is a commonplace of science. So far as we know, the same kind of matter, and the same fundamental laws governing its behaviour, exist in all parts of the cosmos. Temperatures and pressures may and indeed do vary enormously as between earth and sun and even hotter stars, but the basic nucleus and electron constitution, and the basic relations between matter and energy remain the same.

It is in fact true that this world, the earth and all it contains, the planets, the sun, the stars, the whole galactic system, and the universe itself, are the integral parts of one interrelated and interacting whole, every part of which affects every other part. The day of remote particles, individual entities, isolated and separated in their own empty space, is over. Modern physical thought views the cosmos as a field in which forces manifest and interact, and the wave character of the electron extends to the uttermost limits of space.

Fred Hoyle in *Frontiers of Astronomy* expresses the same idea. He says that present-day developments in astronomy are coming to suggest rather insistently that everyday conditions could not persist but for the distant parts of the universe, and that all our ideas of space, time, and geometry would become entirely invalid if the distant parts of the universe were taken away. Or, as seen by the vision of the poet:

> 'All things by immortal power
> Near or Far
> Hiddenly
> To each other linked are,
> That thou canst not stir a flower
> Without troubling a star.'
>
> Francis Thompson, *The Mistress of Vision*

It is in this sense of a visible universe unified with whatever may lie beyond the visible, that the term 'purposeful universe' is used here. Some parts of the vast cosmos are close to man in their influence upon him and he is, thus, more dependent upon them. Others seem infinitely remote, but are none the less significant in their influence on the human spirit, which aspires to the starry heavens and wonders at their infinitude. Man beholds the stars and in our own time is viewing them as new worlds to conquer, either by the delicate extensions of his five bodily senses which are the intricate measuring instruments of science, or by projected journeys to outer space in some atom-propelled space rocket. The vastness and impersonality of the universe may well appall the strongest of men unless the sense of relatedness is experienced. In the human experiences of life, through his contact with that which is near to him, man gains a sense of relationship and unity, and slowly, as his spritual eye develops, he begins to see the wholeness of the vast scheme of nature and becomes aware of himself as a part of a living organic process. As this vision of organic wholeness grows upon him, he comes to realize that he is not only part of, but even in some measure responsible for, the vast cosmic process. It is then that man begins to come of age, and finds his spiritual soul.

CHAPTER 14

INVOLUTION AND EVOLUTION

ALTHOUGH BOTH SCIENCE and theosophy see evolution as the process by which the universe has reached its present stage of becoming, theosophy adds a previous cycle of involution to each cycle of evolution. This concept helps to explain and rationalize the process of evolution to an extent far beyond anything which has so far been suggested by science. The fundamental question which this chapter attempts to discuss is stated by Kenneth Walker in *Only the Silent Hear*, when he asks 'How can anything emerge which was not already present in essence from the beginning?'

Let us see first something of what 'occult' science has to say on the subject. The following is based on the Preface to *Isis Unveiled* (pp. XXX-XXXI) by H. P. Blavatsky. Science ignores and knows nothing about the process of spiritual involution by which the One Power or Force—the Demiurgos of the Universe as it has been called—has become involved with material form. Ancient philosophies, by a strictly logical process of reasoning take as the starting point of the whole process of manifestation which we see before us, the Unseen Spirit, in itself unknowable, but creative in its effects and gradually assuming 'a visible and comprehensible form' to become matter.

Since the most lofty human conception is that of Spiritual Being, from which springs man's highest aspirations, ideals, and conduct, it seems a logical necessity to derive the source of all evolution from a comparable spiritual pole. So with the ancient thinkers we see the process of evolution as preceded by a phase in which pure or essential spiritual force or power becomes more and more conditioned and finally assumes the visible form which we call matter. Within this form are locked up all the future possibilities of evolution, as the spirit unfolds its intrinsic powers. So we arrive at some realization of the cause of evolution itself, for the evolutionary process is the unwinding of the spring which has previously been compressed and wound up. By the limitation of the infinitude of spiritual power into a focus of form the cosmic process develops and brings to expression and realization in actuality the powers and potentialities of the spirit. Such is the origin of the cosmos in the basic philosophy of Hinduism, Buddhism, and our own Western Christian and classical heritage, a view that has become neglected in the empiricism of a modern scientific age.

The *Rig Veda*, the oldest scripture in the world, sets out the process thus: 'In the first age of the Gods, *Being* (the comprehensible Deity) was born from *Non-Being* (whom no intellect can comprehend); after it were born the Regions (the invisible) and from them Uttanapuda (space). From Uttanapuda the earth was born . . . .'

The same metaphysical philosophy has come down in the ancient fragment known as 'The Emerald Tablet of Hermes' — 'As all things were produced by the meditation of One Being, so all things were produced from this one thing by adaptation'.

Involution and *Evolution* are modes of motion or attributes of the Absolute which moves in a dynamic counterpoint continuously and ceaselessly. Time and space emerge from the One, and with their birth come change and growth. The powers and attributes of the spirit are involved into form and so become crystallized within the framework of space and subject to the regular divisions of perpetual duration which men call time. Then, as forms decay and die, with the death of the form comes release of the powers of life. As forms become more complex, as one kingdom succeeds another, the interior qualities of life and spiritual being become increasingly expressed as faculty and function. Life thus achieves ever greater expression in developing new patterns of form. This process of involution and evolution is a continuous one, going on all the time as life here involves into form, and there evolves and transcends the limitation of form. Such is the great cosmic 'Dance of Siva', the Lord of Death and Time. This is the process of continuous creation of modern astronomical theory, a vast spinning dance in which we human beings are immersed and engulfed.

In terms of consciousness, when the spirit, which is essentially eternal and immortal, identifies itself with the transitory and temporal worlds of phenomena, this is involution. *Being* becomes identified with *non-Being*. The very nature of this identification is a bondage to the rhythm of time, and changeableness and impermanence. Gradually the Self, which is free, permanent and beyond change, learns to cease its identification with the phenomenal worlds by an ever increasing self-realization and self-determination as the powers of the spirit become more fully expressed. This is the evolutionary phase of the cosmic process.

In Chapter 1 modern scientific theories of cosmic creation were briefly discussed, and it is evident that what is there outlined is in fact the process of involution, which may now be considered with profit in more detail. When we study the material of which the universe is made, we find that it consists of some one hundred different elements or kinds of atoms, the densest of which are extremely complex in structure. These atoms are storehouses of enormous quantities of energy. By observing the disintegration of the heaviest elements (which are unstable), by harnessing energy to disintegrate lighter atoms, and from the interaction of atoms with radiation, it has been established that the elements of which the material universe is built, have all been formed from the lightest element, hydrogen. This idea of the genesis of the elements from hydrogen, originally put forward by Sir William Crookes as a theory, is now an accepted part of present-day knowledge, but where these hydrogen atoms come from, exactly how or when they are created, remains an open question.

At this point, therefore, we might digress to consider again how the ancient wisdom-tradition explains the creation of the universe. It is held that a supreme Intelligence pervades space everywhere, and by means of its inherent powers creates simple vortices of energy on the innermost planes of nature, these vortices being the simplest prehistoric structures in existence. By combining these vortices into more and more complex groups, new units of greater density are created until it is possible to form a new plane of nature, the finest particles of which are made out of a combination of the densest particles of the preceding plane. This process of densification continues through several phases, gradually working outwards, creating even denser planes of matter until it reaches the simplest forms of physical matter.

By this process, termed involution, enormous forces are built into the physical atom, making it an extremely complex structure of vortices within vortices many times over. As Sir Oliver Lodge said: 'The finger of God stirred the ether and created an atom.'

It is of interest to consider for a moment to what extent, if any, modern scientific discoveries of atomic structures help to confirm this ancient tradition of involution. Up to the middle of the last century the atom was considered as a simple indivisible particle, the smallest fragment of matter that could exist. Then the discovery of radium and its ability to emit large amounts of energy without any noticeable change, gradually led to the present concept of the atom as a complex nucleus composed of sub-atomic particles of various sizes, and having a system of electrons revolving around it roughly like the assembly of planets around the sun. Thus even hydrogen atoms, the lightest of the elements, are not the simplest material particles known. In the natural decay of the heavier elements, and in the products which come from the particle accelerators, many particles lighter than the hydrogen atom emerge, the so-called sub-atomic particles, electrons, protons, mesons, etc. All these sub-atomic particles are forms of energy, and atoms of matter may be regarded as complex centres of energy, 'light itself crystallized and immetallized', to give the appearance of material form.

Although this is as far perhaps as present day scientific knowledge takes us, it is interesting to note that in these particles, much smaller than the smallest atom, we may find confirmation of the ancient tradition of the existence of matter of a finer order than physical, and hinting at the existence of inner and subtler planes of nature. So even the atoms of physical matter itself are now seen to consist of complex systems of energy within the atomic field.

If we accept the hypothesis of the involutionary origin of matter, in which primordial energy, guided by Universal Intelligence, is gradually involved into the systems we call physical atoms, we may see why the universe as a whole from its most distant galaxies to the simplest electron, behaves with apparent intelligence and beauty of design. We may also see that the process we call evolution is the subsequent normal unfolding of growth in a vast universal organism, within which the creation of new atoms and the gradual breaking down of the old, is simply an expression of universal metabolism.

If therefore atomic energy is a localized form of universal life, we should expect that sooner or later it would give rise to a form of matter that would enable the universal purposive consciousness to express itself through physical matter, and that this is in fact what has happened in the course of evolution. Simple chemical materials combine together under the impulse of some kind of energy to form amino acids and proteins, the basic constituent of protoplasm, the substance out of which the living cell is formed. Cells have powers of self-reproduction, locomotion, self-repair and great adaptability to environment giving rise to a considerable degree of self-protection. With these combined characteristics at its disposal, the universal consciousness was able to express itself as a living organism, and exhibit its powers of self-regulation, self-propagation and perpetuation, to some degree, through protoplasm. To the manifestation of the universal purpose as *energy* is added that of *life*.

Thereafter this indwelling life-force, ever pressing outwards for more experience, gradually brought about the evolution of living things from unicellular organisms, through plants and animals, up to the human level, as is generally accepted today, so that man is recognized as the most highly evolved entity we know of in the universe. It is, however, a long story from the development of the living cell to the fully organized human form. Somewhere in the long climb up to man we find the emergence of higher faculties within the purposive driving life-force; those of sentiency, of consciousness and of mind, expressing ever more completely the powers of the indwelling creative force which — for want of a better word — we term God. Sir S. Radhakrishnan, summarizing the ancient Upanishad teaching, says that: 'The cosmic process has assumed the five stages of matter, life, mind, intelligence and bliss. There is an inner direction given to things by reason of their participation in the creative onrush of life', and speaking of man and his place in this scheme: 'Man is a complex multi-dimensional being, including within him different elements of matter, life, consciousness, intelligence and the divine spark'. And again: 'The human being is at the fourth stage of intelligence. He is not master of his acts. He is aware of the universal reality which is operating in the whole scheme. He seems to know matter, life and mind. He has mastered to a large extent the material world, the vital existence and even the obscure workings of mentality but has not yet become the completely illumined consciousness'. (Introductory Essay to the *Bhagavad Gita* translated by Radhakrishnan. Allen & Unwin 1948.)

We may note that the sequence of levels here expressed

by the human being — matter, life, mind, and intelligence — requires a degree of complexity of physical form which has been built up progressively during evolution through the earlier kingdoms of nature. Matter and energy are characteristic of the mineral. Life is characteristic of the plant and lower animal kingdoms. The dawn of sentient consciousness in the higher animals is expressed in simple feeling and thinking processes. The human kingdom alone expresses conscious intelligence. The progression is made possible by a development of physical form, more specifically of a nervous system. Man, in addition to his animal body has the one great difference of a more complex cerebro-spinal organization. The brain, with its co-ordinating areas and regions and complex organization, is the fitting physical instrument for the expression of intelligence. It is a truism that the wider the range of intelligence, the more complex the organization of the upper brain although, as stressed by Radhakrishnan, such intelligence is only partially expressed. Modern man is in fact capable of behaving in the fashion of an animal-man, and may indeed fall lower than any animal in the mentally directed expression of his lowest animal propensities. Furthermore, the complex organization of form endows consciousness with an even more subtle range of exercise of faculty. The static strength and stability of the mineral form of the crystal is subtly moulded into the more plastic organization of cell and plant, which manifests what is usually termed life — biological life — with its characteristics of order, of self-regulation and of reproduction. In the animal kingdom the sentient consciousness of a psychosomatic entity is unfolded. The animal has a rudimentary psyche or soul, as testified by all close students of animal behaviour. Man alone possess the 'I–faculty' which evokes self-consciousness and the greatly increased powers of perception and cognition which are associated with intelligence. How then has this complex evolutionary sequence, up to the partially evolved human being, unfolded?

Since man has evolved from the very dust of which the universe is composed, all his attributes must represent something that is fundamental in the universe itself, for the part can never be greater than the whole. The development of self-consciousness and self-directiveness, with conscious cooperation in the unfolding of the great universal scheme, would appear to be the purpose of the cosmos as far as we can see at present, leading to the harmonious state of being, termed by the Upanishad writers *ananda* or bliss.

The highest qualities of man—his wisdom, love and altruism, his striving for beauty and harmony—must be small reflections of that great univeral consciousness of which he is part. So, by studying man himself as we would study any other natural phenomenon, in all the circumstances under which he expresses the greatest that is in him, we shall be getting nearer to an understanding of that mighty Creative Intelligence which gives the universe its purpose.

Since man has only gained self-consciousness comparatively recently, on emerging from the animal stage, it is natural that at times he should find himself bewildered and working against the progressive evolutionary scheme as he experiments with his newly found mental powers. But here one of the great universal principles comes to his aid—the law of action and reaction. This great law of nature, which in the East is called the Law of Karma, has been explained in such phrases as 'we reap what we have sown', 'as we measure out to others so it is measured out to us'. It is the law of readjustment and harmony, the law of transformation of energy, and it is because of this principle in the manifested universe that we can predict accurately that certain results will always follow certain causes, that they will be proportional, and that the reaction will be exactly what was needed to restore the original balance, and since the universe is a unit, what happens at any part affects the whole, as has been previously emphasized. When Man experiments with life, and in his ignorance strives against the forces of evolution, the unpleasant reactions he brings against himself gradually teach him how to discriminate between that which brings harmony, and that which brings discord. In this way wisdom is slowly born in him.

Coupled with this law of the equality of action and reaction is another great cosmic principle, that of growth by recurrent incarnation in form, the so-called law of reincarnation, as it has been termed in the West. In this process the evolving life discards a form when it has served its purpose and has become an impediment to further growth, and then enters a slightly more complex form in which it can express new powers. Thus, like a reptile casting its skin when the latter restricts its growth, the evolving life continues to discard old forms and enter new and more suitable ones throughout the whole course of evolution. In this way consciousness and form are seen to evolve side by side, the one making full use of the other at all stages, in a kind of cosmic symbiosis.

This fundamental principle of cyclic incarnation of life in form has been frequently misunderstood in the West, but when recognized as occurring in all Kingdoms of Nature it explains why she is so lavish in her production and destruction of forms in the process of growth. It is then realized that the in-dwelling life of each form is not destroyed, but simply leaves one form for another more suitable, while the form-making energies of Nature are themselves gaining in experience as they build more and more complex structures.

And so we gradually come to recognize the Universe as the manifestation of a mighty creative intelligence expressing itself not only in ever-changing forms, but also in those attributes of love, compassion, wisdom, and beauty, which constantly shine through every creature, for those who have eyes to see.

CHAPTER 15

THE FORMATIVE IMPULSES IN EVOLUTION

IN THIS CHAPTER the foundations of the purposive drive for life itself within the very substance of the material world will be examined. Present day fears and uncertainties are to some extent the result of our incomplete acceptance of the change that has taken place in the materialistic patterns which dominated western thought in the latter part of the nineteenth century. Our failure to grasp the full consequence of the view of nature as dynamism and energy, which is the result of present day scientific discovery, is at least one of the major obstacles to the establishment of a spiritual world civilization.

We have looked at the constitution of the material or physical world, and considered the process of its formation and evolution and the emergence of living entities therein. We must now discuss what is the motive power, the inner drive, of biological evolution. Whence arises, for instance, the 'will to live', which drives all animate nature to experience, to increase life, to propagate? In Chapter 2 it is suggested that this drive is the outcome of a divine or spiritual creative principle which is ever seeking freedom and expansion and self-expression through material forms. In this chapter we shall discuss in more detail how this spiritual principle may operate in the building of forms for its own fuller expression.

Many biologists still believe that living forms emerged gradually, and in a sense automatically, from mineral matter. Climatic conditions on the earth when only mineral forms were in existence could lead to the formation of low concentrations of amino acids and other organic molecules capable of providing nourishment for exceedingly primitive unicellular organisms. From this unpromising beginning, the rich variety of plant, animal, and finally human forms was supposed to have arisen by gradual inevitable steps, through the operations of genetic inheritance, chance mutations and natural selection.

Compressed thus into three sentences, the whole idea seems ludicrous, as if a watch should assemble itself, given long enough, from a pile of scrap metal. As the complexities of biological organization and function were unravelled, and more recently also the underlying molecular intricacies of biochemical systems, so has this theory become increasingly untenable. The verdict 'reductio ad absurdum' might have been passed long since, if an alternative hypothesis acceptable to materialistically minded scientists had been available. Characteristically, the dying theory is being vigorously championed by the diehards, but the relentless force of their own logical minds is compelling many scientists to seek less materialistic hypotheses, because no completely mechanistic one seems competent to explain the facts. Several biologists in recent years have put forward the concept of an inherent purpose in living organisms, that has guided evolution, and still guides the development of the organism from the germ-cell to the adult form, keeps it functioning through its life-span and controls the repair of damaged parts. Such a concept is not so different from that of the 'One Life' pervading the whole creation.*

It may now be possible to set down briefly an hypothesis concerning the origin and development of living forms that would be acceptable to at least a few biologists, and also to students of theosophy. Naturally one must not expect the same terms to be used by the two schools of thought, for each has come along its own path

---

* See also *Intelligence Came First*, published by the Theosophical Publishing House in America, Wheaton, IL, 1975.

on to common ground; rather must one penetrate through the two terminologies to consider whether the ideas behind them are not very much the same. Terminology becomes especially confusing when the same word is used in different senses by the two schools, and this is so with the fundamental word 'life'. To the scientist life *is* that which distinguishes vegetable and animal forms from mineral; the mineral world is non-living, *by definition*, and it is meaningless to suggest that this is not so. The theosophist on the other hand regards the One Life as manifesting to varying degrees in all material forms, including the mineral. Even so, the transition to organic forms marks such an increment in the freedom of the life-force that a new 'Life Wave' is postulated, so the difference is still only terminological. It is perhaps best to give the word 'Life' a capital to convey the theosophical usage. Life in that sense is one of the supreme fundamentals, without beginning and without end; it is entirely beyond analysis by the human mind. However, this is not to say that the scientist is wrong to speculate about the origins of life in *his* sense of the word.

Initially the earth carried no organic physical forms. However, the mineral matter was in a far from disorganized condition, turbulent though the surface of the planet may have been in those distant times. The ultimate particles of physical matter exhibit a capacity for ordered aggregation. On this point there is profound agreement; whether one looks to the atomic theories of science, or to the clairvoyant observations recorded in occult chemistry, the chemical atoms appear as organized entities, each one of a kind like every other, and displaying orderly sequence from element to element. This patterning is continued at the molecular level, and again when molecules aggregate into crystals. These patterns can only be revealed by techniques like X-ray crystallography, but the shapes of individual crystals and the patterns made by their aggregates, such as ice crystals, may be visible to the naked eye.

With all this in mind, there is a tendency to suggest that

The Formative Impulses in Evolution 131

matter had already inherent within it the capacity for the more complex organization of organic forms, so that these arose 'spontaneously' when conditions on the cooling earth's crust became suitable. Recent experiments have shown that amino acids and other organic molecules that could conceivably combine into a primitive protoplasm, could have arisen by lightning discharges through mixtures of gases likely to be present in the earth's atmosphere at the time. No scientist can suggest in detail how such material could become 'alive' in the biological sense, or why spontaneous generation never occurs now. In some ways, too, death is harder to explain than life; why does organization as an entity suddenly cease, sometimes without any evident external cause? There is an area of ignorance here, admitted by science and inaccessible to investigation by present techniques. In any event, it is hardly possible to pretend that there could have been smooth continuous transition from mineral to vegetable and animal forms, however primitive. There is a qualitative difference between the ordered aggregation of the one, and the purposive ordered integration of the other, which is indeed recognized by the very word 'living' applied to biological forms. If we are to concede purpose in the scheme of things, as we surely must, then there was a great leap forward when life appeared. A crystal may be said to feed and grow, but it feeds upon the same single substance of which it is made, and it grows by accretion, not by assimilation of selected portions of a mixture of foodstuffs and their chemical modification into protoplasm. A crystal may serve as a nucleus for the growth by accretion of a new crystal, but this is quite different from the division of the contents of the living cell, to form a replicate daughter-cell. A living organism is a stable system, but not a closed one. If an organism is confined, and its life is investigated, the Second Law of Thermodynamics would indicate that, while the total energy remains unchanged during the organism's process of eating, doing work and discarding refuse, the entropy must increase. This in fact does not occur. Assuming an adult organism to maintain an approximately stable state, it may be said that the entropy of food is smaller than entropy of waste, or, that an organism feeds on negative entropy. That localized sinks of negative entropy are to be found in nature does not seem a reasonable deduction from existing materialistic postulates, and suggests that some additional factor has been ignored.

We find ourselves in the situation in which the biochemist, with the exercise of much thought, care and purposive activity, together with the use of elaborate and intricate equipment, can duplicate in his laboratory a few of the simpler processes known to be performed at a considerably higher level of complexity by a single microscopic living cell—the activities of the cell being ascribed to hazard and chance!

Experiments with radioactive isotopes have shown that there is a constant turnover in the material constituents of living organisms. Matter flows endlessly through the cells, not only for replacement of worn-out substance, but as a natural concomitant of being alive. It is therefore difficult to ascribe the known characteristics of living organisms to their constituent matter, because this is only transitorily present. The *design* of a living body is all that continues in being.

We, as theosophists, need to put the utmost emphasis on the discovery that the life and form of all living cells and hence of all living beings, including humanity, is controlled by a non-material field of force which allows physical matter to enter and leave this field with ease, but which controls it while it is in the field. This is especially significant in connection with human consciousness. Although the matter of our bodies is constantly changing (in some parts very frequently) we retain our identity of form, and our continuity of consciousness. Memory, and the sense of self-hood must depend on the non-material field of force or framework and not solely on the brain. This framework is more permanent than the body. It may even outlast it when the body dies. Man is therefore a non-material being using a body whose parts are in a state of constant change.

A living organism proceeds on its way from seed or embryo to adult as if it had an aim. If it is thwarted, starved, or mutilated, it does its best to repair the damage and come as close as it can to the archetypal form of its species. Reflections such as these have caused biologists to postulate a capacity for goal seeking as an inherent property of protoplasm. In his book *The Biology of the Spirit*, Dr. Sinnott says: Directedness is the most characteristic feature of a living thing. Protoplasm does not make formless structures, it builds organisms.' 'A living thing is one organized self-regulating system — an organism.' 'This integrating and directive control so resembles the directiveness of man's behaviour which we call mental or psychic activity as to suggest the exciting possibility that the two may be expressions of the same underlying biological process.' This accords well with the idea of the unity of all life, though the theosophist would extend its scope beyond the biological manifestations. However, biologists are not ready to postulate any disembodied reincarnating entities, nor even a generalized Life-force capable of existing apart from the physical forms. Whatever one may believe on these matters, there is no difficulty in agreeing that a capacity for organization is inherent in matter and to a greater degree in protoplasm. This may be thought inadequate, but it is an acceptable way of generalizing observations, especially when the capacity for 'goal-seeking' is added to the properties of protoplasm. A goal implies a pre-determined objective, and goal-seeking implies purposeful endeavours towards that objective. This surely, is none other than the age-old doctrine of Archetypes, expressed in new terms and now derived from patient observation and experiment.

Gradually it was accepted that the power to vary was within, that the organism itself has unconscius control over its own variations, and is not forced into specific development by outside forces. In other words a race has an unconscious ideal and develops towards it. R. C. Johnson in *The Imprisoned Splendour* has expressed this: 'From time to time some great challenge seems to surge out of the vast unconscious of nature—"to capture a new area of life". As though in response we see experiments taking place and novelties constantly arising. Some seem to last a few million years and then fail. Others succeed and from these again new experiments arise. . . .'

Graham Cannon sums up the situation as he sees it in his book, *The Evolution of Living Things:* 'All the evidence points inexorably in one direction: to the conception of some guiding force within the organism which controls and guides its evolution, not by haphazard changes but by selected modifications'. This is the perfecting tendency of Aristotle, the good genius of Bergson, the entelechy of Driesch, the harmonious control of Waddington, the balanced evolution and organismal control of Cannon himself, and the goal-seeking of Sinnott.

We may also quote E. S. Russell in *The Directiveness of Organic Activities:* 'We recognize the fact that organic activities . . . show . . . directiveness, persistence, and adaptability. . . . Human directiveness and purposiveness . . . are a specialized development of the directiveness and purposiveness inherent in life'. 'In living organisms the final cause of development is an immanent internal directiveness towards a goal.'

It may be suggested that the concept of goal-seeking implies an Intelligence which set the goals and implanted in protoplasm the urge to seek one or another of them and so develop eventually into a tree or an animal. On this, science has nothing to say becasuse it is beyond the reach of the scientific method. A few scientists, nonetheless, speaking as philosophers expressing their private faiths, have indeed assented to some such ideas, which are certainly not excluded by the scientific evidence. One might go further and suggest that anything so dull-witted as protoplasm could hardly have maintained through aeons of slow change that initially-implanted drive towards, say, a forest oak; would not such an extensive and remarkable transformation require external guidance, not just initially but continuously, or at least repeatedly? Perhaps it will not be so many years ahead before biologists will be prepared to accept the testimony of the great religions for the Creative Gods, the Elohim of the book of Genesis, or the evidence of psychics for the existence of the building Devas. As Johnson says in *The Imprisoned Splendour:* 'The process bears the general character of intelligent research, of countless experiments directed towards a goal'.

But most biologists still accept *chance* as the guiding principle — if principle it can be called. Individual variations are usually brought about by haphazard breaking of chromosomes and crossing over of parts in the process of recombination following the union of male and female germ-cells. Variations also occasionally arise by mutation, the alteration of a gene that can be brought about by radiation, for example, by cosmic rays. These observations are interpreted on the principles of natural selection, popularized by Darwin and his exponents, i.e. on the idea that in the struggle for existence that goes on throughout nature, the fittest organisms are those that survive in the long run. Thus a chance variation would tend to persist and become established only if it enhanced the survival value of the organisms. In this fashion the 'principles' of chance and ruthlessness have become enshrined as articles of faith, as the accepted laws of evolution. So it came to be seriously believed that all the beauty, perfection, and diversity in nature arose in this totally mechanistic way, and owed no more to any guiding principle or initial design than does the winning number at a roulette table. These ideas have held sway for many years, perhaps because they matched the general climate of opinion and the intellectual arrogance of the Victorian era. From time to time, however, scientists have perceived intuitively the inadequacy of these bleak concepts, and have sought a way out of the impasse, a way to dethrone these hypotheses that seemed so soundly based on observation.

Schrödinger for example in his little book *What is Life*, made a tentative start. He suggested that the all-important genes, which may be molecules of 1,000 or less atoms, are small enough to escape from complete control by the laws of physics, based as these are on the statistical average behaviour of large numbers of particles in random movement. That is to say, a gene may be modified by a single random event of sufficient intensity and the effect need not be cancelled out by other events causing the reverse change, as would happen on average to a large concourse of molecules. Statistical theory produces order from disorder, but living organisms convert one ordered form into another; 'an organism feeds on negative entropy' which may be loosely defined as orderliness. Schrödinger makes out that this kind of reasoning led him to mysticism, though one may suspect here a reversal of the true order of events. We are led to take the idea a stage further by suggesting that if one substitutes for 'chance', active intervention by an unseen intelligence, then the observable result would be the same.

Thus escape from the thraldom of chance seems possible by suggesting that the selection is not so automatic as has been assumed. Deliberate selection has long been exercised by man in his breeding programmes, and it would be hard to prove that nature in the wild does not also exercise some choice by protecting entities destined for survival. This could well be the manner in which the unseen intelligences, the Devas, operate. The idea does not seem to have been suggested so concisely by any biologist, and probably none of them would be willing to subscribe to it, yet nevertheless it does appear to be implicit in some of the recent hypotheses.

In any event, the combination of chance plus natural selection as an explanation of the progress of evolution must surely be rejected or greatly modified, because it simply does not work. One of the main objections is that the many changes involved in the development of a new organ, for example, or in the transition from life in the sea to life on land, are closely co-ordinated and integrated. If genes were in control of the situation then a whole series of mutations would have to occur, in a fairly closely defined sequence, to complete the transformation. Now the probability of all this happening by a series of lucky chances is so remote that geneticists have had to think up a way round the difficulty. This is where Waddington's hypothesis of harmonious control comes in; it seems to envisage a mechanism that makes such things happen the right way—a kind of super-gene that keeps the rest in order.

Another fundamental objection is that chromosomes represent a highly complex and sophisticated mechanism of variation. Evolutionary changes were presumably taking place successfully during millions of years before ever nature got around to inventing genes. This seems utterly obvious once it is pointed out. Graham Cannon goes further by suggesting that Mendelian inheritance can have played no part in the major steps in evolution. He claims that this mechanism can only produce modifications of what is already established; it does not seem competent to produce, say, a new organ; at any rate all that can be demonstrated to be under Mendelian control are the relatively trivial details like colour and pattern.

Although micro-organisms can and do mutate, in general, sexual reproduction is an essential concomitant of gene inheritance. Graham Cannon goes so far as to wonder why nature bothered with sex at all, when she was getting along so nicely with division or budding as a means of reproduction. He does point out that the inter-species sterility that is part of the gene process may have been needed to maintain the separate species. One might venture further and suggest that without sexual reproduction the full potentialities of the vegetable, animal, and especially the human kingdom could not be achieved.

As any gardener knows, plants resulting from vegetable reproduction grow to be exactly like the parent, whereas seedlings (resulting necessarily from sexual union) show some individual variation, even when the strain is supposed to be fixed. As the number of genes increases so do the possibilities for individual uniqueness, to say nothing of the educative experiences associated with sexual reproduction among the higher animals and humanity.

The second evolutionary mechanism that Cannon postulates as predating Mendelian inheritance and presumably still continuing alongside it, can at present only be formulated in rather vague terms; 'Organismal control is there to see that the organism is in the right state at the right time'. Both Cannon and Sinnott regard such a mechanism as a quite inescapable deduction from unbiased observation of living organisms.

A weakness of the gene hypothesis is that major changes would require the slender chance of many successive mutations occurring in the right order and the right direction. Organismal control or balanced evolution, it is supposed, can contrive gradual co-ordinated changes. However, either hypthesis seems to involve many generations of organisms in various stages of transition from the old form to the new; but these half-way creatures would almost certainly be eliminated in the struggle for existence, because they would be handicapped, being less well fitted to survive than either the old form or the new one yet to come. This consideration has yet to be squarely faced, but it does seem that 'survival of the fittest' may also have to be overthrown as an evolutionary mechanism, at least in its original form. It appears that nature must have found some way to protect these transitional forms from premature destruction. If this should be accepted, it would constitute another major example of purpose within the scheme of evolution. Moreover in this manifestation it can hardly be regarded as a property inherent in protoplasm, for it is required to operate in a broader field, covering presumably thousands of creatures during thousands of years. To suggest that all this might have taken place in some quiet underpopulated corner of the earth where the struggle for existence was temporarily abated, is really only to express in other language the hypothesis of a protecting influence, superimposed upon a guiding influence. If one feels compelled to accept either of these, or both, then there seems no logical escape from the concept of a great purpose behind creation.

One may be tempted to question whether these niceties really matter to a humanity harassed by the threat of atomic extermination. During the last century men have on the whole felt constrained by some inner power to behave decently, in spite of a scientific philosophy that provided no basis for such altruism. But such a dichotomy between reason and intuition has a disintegrating effect which is all too evident in the world today. The old principle of the 'survival of the fittest' is little more than a statement that those organisms which survive are those organisms which have survived! Attention given to ruthlessness and conflict in nature blurs the harmonizing and balancing factors which are equally present.

The recognition that purpose in nature may logically be admitted is overdue and permits a sense of a rightness in things, long ago glimpsed intuitively. The world becomes at once a more hospitable place to live in than one which seemed to be founded on the laws of the gaming table. The mystic in most of us receives fresh inspiration through a sense of unity with that purpose. One can take more joy from a sense of the staggering intricacy, the almost unbearably beautiful fitness of the parts of living things'. (W. W. Ballard in *Science* for April 25, 1958). These ideas come close to that of the 'One Life' pervading the whole universe, which is implicit in Eastern philosophies and in modern theosophy and too in Christianity, if one may extend a little the concept of 'Him in whom we live and move and have our being'. If man is to become whole again, he *needs* a philosophy that includes and gives form to the purposiveness he senses in the scheme of things. As Graham Cannon says in *The Evolution of Living Things*, 'Evolution . . . is the result of something central which is inherent in all living things. It is not simply a succession of lucky chances'. This central something is, it is suggested, the Spirit, the One Life, the God Immanent, whose involution into every atom was discussed in Chapter 2, and whose ceaseless urge to freedom, to self-determination and self-realization, is the motive power of the whole stupendous process of cosmic evolution.

---

* See also our later book *Intelligence Came First*, mentioned in the Foreword.

CHAPTER 16

MAN AND SOCIETY—A UNIFIED VIEW

IT IS A COMMONPLACE of modern scientific thinking that evolution has resulted in the emergence of three distinct levels of objective reality or 'grades of significance'. These are termed respectively the inorganic or pre-living, the organic or biological, and the psycho-social or human levels. These have been referred to by several writers (see especially *Evolution in Action* by Sir Julian Huxley). When first discussed by the American scientist, Prof. Birkhoff, before the American Association for the Advancement of Science in 1938, five levels were mentioned, namely: mathematical, inorganic, organic, psychological and social. It has since been realized that the two levels, psychological and social, although at first sight distinct, in reality are one level only—the psycho-social. This is because man is essentially a social being, and from the earliest times has organized himself into societies. Man differs fundamentally from the animals in that he has developed a mind capable of abstact thought, of planning for the future, and what is most important, of seeing himself as a member of and in relation to a social group. It can be said indeed that man is only truly human in relationship with his fellows. It is extremely unlikely that any such sentiments actuate the animals. It is true that in a few exceptional cases, e.g. bees and ants, a co-operative, communistic state of society exists, but in general each animal is entirely individualistic, and apart from the breeding season and the short time that the young are in need of parental care, there is probably little real co-operation and practically no social consciousness manifested among the animals.

It is the social sense which gives to the Western world its idea of good citizenship, and although each citizen is playing his own hand largely for the benefit of himself and his family, yet the game is played in accordance with well-defined rules, and transgression of the rules makes the criminal. That there is, and must be, a stage beyond this 'enlightened self-interest' is evident, for increased technological knowledge makes the struggle for self-preservation more deadly, and unless human society is to be entirely disrupted from within itself, it is evident that altruism must gradually replace egotism. This is indeed being widely recognized. The following from *Elements of Mathematical Biology* by A. J. Botha supports this view: 'While the human species ... has become organized into a social whole, the motivation which keeps it going ... continues to be individualistic in type. The competitive element is not without salutary action ... but ... may there not be evolved in due course a superior system which shall secure the interest of the community more directly and with less internal friction ... (and) ease the conflict between man and man and between man and the world at large?' (Quoted in *Main Currents in Modern Thought*, March 1958.)

Now man's psychological, ethical and spiritual development is akin to that recapitulated by every child as it passes from birth to the adult stage of life, and the soul that is the inner man becomes progressively expressed. A baby of six months has begun to be aware of his relationship with his senses and his surroundings. In watching a child of this age one can see the dawning experience of sensory perception and the registration of contacts with the environment. In much the same way does the human race, in its evolutionary progress, become aware of the world in which it lives. At first, in primitive races, this realization tends to be receptive, egocentric and self-related, lacking precision and detail. This phase is that of a child who seeks to master his senses, makes big movements, and cannot make precision actions, until he learns a truer relationship between himself, his senses and his surroundings. So the first task of evolving man was to bring the physical environment into relation with his inner world of perception through the medium of the senses. As some measure of success is achieved in this, a growing child enters upon the next phase, the dawning emotional period of adolescent youth. On the wider scale also, young humanity, feeling the majestic rhythms of the material universe and all that is comprised in the idea of 'Nature', has expressed its feelings, first in propitiation of the gods of nature, and then in the arts and true religious aspiration. Through wonder and feeling man discovers a creator and a purpose behind the universe, which he terms God. Another phase then supervenes and the mind becomes the dominating mode of conscious experience, and the chief channel for communicating with his fellow men; and by investigating nature the growing adult obtains more comprehension and understanding of the world in which he lives.

Humanity, like the infant, has thus experienced the stages in which consciousness is focused in sensory perception, in feeling and in the thought processes of the mind. Although these phases overlap in their unfoldment, they arise in this ascending order. Primitive man with his direct sensory contacts with nature, physically skilled, but primitive in feeling and superstitious in belief, lies at the heart of each one of us. Superposed on this, however, we have the phases of later human civilizations, in which the mind develops in its early stages a very clear cut manner, as it deals with the concrete and immediate needs of the personal life, later extending its ranges to the more abstract and perceptive faculties. We have arrived by this path at our present-day age, that of science, with its intellectual complexities and knowledge. These are so intricate that they may well lead to confusion and loss of purpose and aim, unless a deeper vision is attained.

The development of the individual human being in the light of the theory of reincarnation affords yet a further useful parallel for the understanding of the wider human problem. During his first twenty or so years of life each person, being the expression of a re-incarnating ego, recapitulates his own growth in earlier lives, and brings through into physical expression already evolved faculties of the soul, following the above sequence of powers of the body, the emotions and the mind. As he reaches mature age he begins to break new ground. Some individuals progress in this manner further than others, and so we may perceive almost the whole range of development from the primitive to the genius. We may look upon the diverse races, national groups and societies within the world as representative of the various levels of unfoldment of spiritual potentiality. Indeed this principle is implicit when the symbol of the family is applied to the human race. Much of the ferment today arises from the rapid technological advances which are stimulating backward racial and national groups (the younger children) to unfold their intellectual faculties at an accelerated rate. Before humanity can go ahead and build a world civilization these different cultures must be integrated, and their mutual relationships understood, in the light of the realization that they are diverse expressions of the evolving human spirit.

In the past, taking a long-time view, civilizations seem to have succeeded one another in an orderly pattern, each maintaining its own characteristics and individual stamp. Today, in part owing to the rapidity of communications, they seem to be intermingled, so that on the world stage of the present era we find racial groups and civilizations at many different levels of development, all pressing closely upon each other, intermixing and exchanging habits, customs and ideas. Amid the welter of conflicting interests which emerge, the world problem is to find some way by which these diverse groups may live happily one with another. This phase is comparable to that of a large growing-up family, in which the various individuals have to find some mode of adjustment within the family unit. At the age of sixteen to twenty or so, a young person entering an adult life has to submit to a socialization process, and make progress in all aspects of social communication with his fellow men. In a similar manner human civilization, now entering upon an adult stage, is embarking upon a socialization process, in which national and racial groups have to evolve in common a wider social community. Through technical advances which have made the world remarkably small, and have speeded up communications in an incomparably rapid way, man is faced with the task of integrating differing racial groups into a world civilization. The various organizations with their portentous letters, UNESCO, WHO, UNICEF and others, all represent the beginnings of human effort to co-ordinate social activities on a world scale. Such a process is ultimately as inevitable as that by which water flows from a mountain spring down to the plains below. As the needs and claims of society impinge upon a growing adolescent, living his own thoughts in his own private world, and force him into ever widening relationships, so humanity having arrived at the threshold of adult life must socialize its activities or perish through the massive powers which have been developed by science. Technical and economic powers and the vast resources of human knowledge are already developed to an extent that makes a world civilization a possibility. All that is wanting is the will to co-operate and to use those powers for the good of the whole world and in the service of all men.

This vision of a world unity to be established by mankind appears hopelessly impracticable and unrealizable so long as each man feels himself limited and circumscribed by only one life of a few score years. Only the most altruistic and visionary are content to work for the good of humanity in the abstract, although it is worthy of note that there are the altruists who do put their powers at the service of their fellow men without holding any belief in personal survival after death. For many, however, some deeper insight into the meaning of life is required. The concept of the progressive re-embodiment of the human spirit in form, the process termed rebirth or reincarnation, gives meaning and hope to individual human life. Most thoughtful people today, except those of an entirely materialistic habit of thought, envisage some form of continuity for the human spirit after death, and the necessity for its continuing education and growth in spirituality. The cyclic rhythms of the universe all follow a repetitive pattern which makes it at least feasible that the human soul may have recurrent periods of incarnation in bodily form. A belief in the ultimate perfectibility of the human soul, that 'far off divine event to which the whole creation moves', draws strength from the idea of cyclic incarnation of the human being in physical forms, that he may develop progressively the faculties of the true spiritual man. It is a commonplace of biological evolution that forms become increasingly more complex in the successive kingdoms of nature. As the functions of the interior life become more fully expressed in more elaborate forms, the simpler entity seems almost to be re-embodied in the later one. The plant form, based upon the mineral pattern in its basic structure, is in one sense a reincarnation of that mineral, with powers of flexibility and growth superposed upon the stable mineral form. The animal form, and later that of the human being bear a similar relationship to the kingdoms below them, and so in the various kingdoms of nature, we may see all the stages of unfoldment upon the world stage together.

This process of cyclic unfoldment from one kingdom to another is essentially the process of evolution, which is thus the expression of the cyclic incarnating process. The concept of human reincarnation at its best is, however, more than biological evolution, for it involves a progressive entering into form of a life-energy from a permanent centre of individuality, which absorbs the distilled experiences of incarnation, and so over long ages expresses increasingly its potentialities as a creative living entity. Such, briefly, is the cyclic pattern of the individual reincarnation process, and since cyclic rhythms appear to be one of the most general of the laws of nature, we may look beyond the individual and see the same rhythms in the wider field of society.

The growth of social units, races, nations and civilizations proceeds in a similar manner. Each civilization passes through a cycle of birth, of life and of death, as it struggles into being, develops into maturity, and finally tends to pass into decay when it becomes crystallized and fixed in a too static pattern. The impulse within it seems to reach a certain stage and then a new urge toward human perfectibility manifests in another group. This has up to our times led to the rise and fall of mighty civilizations, one succeeding another as the earlier is weakened by internal dissensions, attack from without, or a blend of both.

There seems no reason to expect any different pattern for the so-called modern age of science, although our deeper insight into the pattern and purpose of the life-process itself may enable us to identify ourselves with the continuity of the living process rather than with the static form. So too civilization may pass into a more spiritual phase by evolution rather than by revolution. Tragic though the fall of a civilization may seem, it can be recognized as the natural consequence of cyclic law, and the death of its form can be accepted in this field of human endeavour as in all other living processes. Conflict, tension, and indeed the dissolution of much loved institutions may thus be seen as a part of the pattern and mechanism of the process by which human beings are spurred on to effort and deeper self-realization.

It is an obvious experience of our times that the world soul is emerging through conflict, and that progress results from tension and the threat of disaster. Totalitarianism, communism, and the appalling threat of the hydrogen bomb have all played their part in bringing to birth world consciousness. In the economic fields the needs of the poverty-stricken millions in Asia and the near East, are brought on to the doorstep of the more materially advanced western technological countries, as an immediate personal problem to be solved. In the political field the tension produced by the urge to freedom and independence of colonial and depressed racial groups, leads to experiments in progressive forms of governments undreamed of a generation or so ago. Mankind in our times is unconsciously seeking to express its inherent unity in what might be termed a social planetary consciousness, and to develop a vision of the purpose of life itself as distinct from the processes of living. Only this can give security and the dignity of life-fulfillment which are sorely needed in place of fear and frustration.

CHAPTER 17

MAN'S TRUE ROLE IN THE UNIVERSE

OTHER CHAPTERS in this transaction have discussed the unity of man with nature, and have pointed out that man, the macrocosm (the large universe). This implies that man is in some sense an epitome, a reflection, or a miniature copy of the cosmos. Man is also said to be made in the copy of the cosmos. Man is also said to be made in the image of God, and Annie Besant, writing many years ago, said 'The essence of theosophy is the fact that man, himself a spark of the divine, can know the divinity whose life he shares'. Mystics, yogis, seers have throughout the ages affirmed that the essential spiritual Self of man is in fact one with the Self of the universe — one, that is, with God.

Now there is a fundamental difference between the view of cosmic evolution held by most scientists and that expounded in theosophical literature. In *Doubt and Certainty in Science*, J. Z. Young writes 'Man is a very recent product of evolution and it is tempting to look upon the whole process as leading in his direction. This attitude we must resolutely dismiss'. It is essential to the scientific method to examine and record the facts, to study and if possible to explain the evolutionary process without assuming that it is leading anywhere. Such an assumption would inevitably distort the purely objective view which is the scientific ideal, making it anthropomorphic and teleological. To a theosophist on the other hand, attempting as he does to take a widely synthetic — or better, an intuitive — view, the whole of nature *is* anthropomorphic and teleological. Teleological means working with a purpose, which purpose need not however be assumed to be foreseen or predetermined as otherwise the scheme becomes purely deterministic and free-will is negated. It is true that some thinkers have denied free-will, but whatever the mind may say, the fact remains that the conviction of free-will is one of the deepest intuitions of the human heart without which living would be meaningless. Even those who most strenuously deny free-will to the human being and purpose to the universe still act as if they themselves were free and their lives purposeful. Purpose implies a plan and the only alternative to a plan is pure haphazardness which again is unthinkable. Even though large-scale phenomena are regarded as the statistical result of innumerable apparently random events, it is still impossible to believe that any one of these events has no cause.

Man's view of the universe must necessarily be anthropomorphic because in the last analysis he can only be aware of himself and his own reactions, which self he inevitably projects on to his environment. This does not of course negate the value of a study of external nature and of man's place in the universe — indeed it rather adds to its value.

Looking back at the course of evolution in the light of previous involution as discussed in Chapter 2 it becomes evident that the chief result has been the gradual attainment by life of an ever-increasing measure of freedom. This movement towards freedom of the spirit through the increasing sensitivity and reponsiveness of the bodies is evidently one of the purposes of the evolutionary process.

It is of utmost importance to realize that any attempt to understand a particular stage in the process by consideration of the past alone is doomed to failure. The present is determined equally by the future as by the past, and each evolutionary level can only be comprehended in the light of future development. Sir Julian Huxley has said that man—because of his creative intelligence—is no longer pushed forward by the past only, but is also drawn forward by the future. This is true of man but it is also true of the rest of nature. In the case of man the forward movement is conscious and, in relation to anything which has gone before, is incomparably rapid. The whole process can, however, best be regarded as leading to man, just as man himself can best be understood if he is regarded as a stage on the way to a level beyond. This can be realized for oneself by a deep introspective examination of one's own consciousness, and an attempt to assess one's motives. While many beliefs and activities are undoubtedly motivated by sub-rational urges in the past, some also are from a supra-rational region of the psyche which is leading to the future.

Science and theosophy agree in seeing man as the result of a very long evolutionary process leading upwards through the inorganic and biological levels to the psycho-social or human level. Many scientific workers are beginning to see a unitary principle at all levels of outer reality. One would in no sense denigrate the magnificent work of science, but the unity which is being perceived is not only physico-chemical, but is surely the expression of the unity of the in-dwelling life. Theosophy, while agreeing with the concept of evolutionary levels sees them largely in terms of stages in the unfoldment of Life. It does not, however, teach that a stone becomes a plant, that a plant becomes an animal, and an animal a man. The theosophical concept is that the natural order is the result of a 'Life-wave' descending during the involutionary phase, and then ascending through kingdom after kingdom, gathering experience as it goes. This is what is meant by the statement of H. P. Blavatsky that every living entity either is man, has been man, or will be man.

It cannot be too strongly stressed that man is unique among the creatures, that he is not merely a species of higher animal. Man has will and individuality in an entirely new way, in which he differs from the animals and, according to occult tradition, from the devas and nature spirits also. Man is the crux of the evolutionary process. This is beautifully expressed in a quotation from *Protean Man* by D. N. Dunlop who writes: 'for creation from the first has been a continual effort to put forth the human form. Mineral, vegetable, and animal forms, nay atmospheres, planets and suns are nothing else than so many means and tendencies to man, on differing stages of his transit. He stands on the pyramid of being, linked with all below as the form to which they all aspire. Man is the head and heart of nature. Creation is the coming and becoming of man. The world is because he is. The reason of everything it contains is written in the book of human nature. He finds that reason physiologically in his body and spiritually in his soul'. In theosophical literature we find it stated that 'physical nature keeps pace with the physiological as well as the spiritual man in her cyclic career' *(Mahatma Letters to A. P. Sinnett*, p. 47). We tend to think of civilization being advanced or retarded by man's discoveries and inventions. It is even more true to think that man discovers in nature that which he needs for his own next step.

The theosophical teaching that involution into matter has occurred through three manifestations of the Life-wave, the Three Outpourings, can be of assistance in attempting to understand man's place in the scheme. The first drive into form or outpouring from the Third Person of the Trinity, the Third Logos or Creator awakens and fertilizes the virgin matter of space, forming in the continuum innumerable discrete structural units or centres. These, under the influence of a second wave or outpouring from the Second Logos, the 'Divine Weaver', are built into patterns and forms ensouled by the Logoic Life, forms which evolve as minerals, plants, and animals. Suitable vehicles for life expression having been thus evolved from below (but always, be it noted, under the stimulus of the previously involved indwelling life from above), the Third Life wave or outpouring from the First Logos now occurs. This is like the release of tension in a lightning flash from sky to earth, and by it the human monad is connected with the lower vehicles and the true man is formed. Thus there comes into existence a being in whom *manas*, the principle of intelligence, forms the link between highest spirit and lowest matter. This *manasic* principle embodied in the causal body is the basis of the immortal individual who persists from life to life. It is because his psyche is irradiated, however dimly, by the light of the First Logos, that man has a persistent intuition of free will; that his creative intelligence is capable of long-term planning; and that he has individual responsibility for the results of his thoughts and actions. In the *Mahatma Letters* it is stated that 'Up to *man* "life" has no responsibility in whatever form' (p. 74); and again '. . . man, whose intelligence makes him the one free agent in Nature' (p. 57). It is because the gamut of his being ranges from highest spirit to lowest matter that man is the microcosm; and because he embodies a ray of each of the three Logoi that he is 'made in the image of God.'

Having all the divine attributes in embryo, it becomes clear that man's responsibility for the future, both of himself and of his planet the earth, is very real. This is why, to many theosophists, the great work is the enlightenment of humanity, beside which all other activities pale into comparative insignificance. For the individual, the true human task is that of spiritual Self-realization, of ceasing to identify himself with his personal cravings and anxieties, and learning to act from his own spiritual centre, which is equally the spiritual Self of all men. In the mass, man has as yet hardly begun this realization, and tends to identify himself almost exclusively with his personality, with the satisfaction of the needs of his body, and the desires of his psyche. The mind, which is the unique human principle, is the means by which man's progress is accomplished and his influence exercised. Dual in its expression it is poised as it were on a knife-edge between the desires of the personality and the will of the spirit, and it is through the mind in full physical plane consciousness that man will find himself, for as stated in *The Crest Jewel of Wisdom*, '. . . Mind is the cause of his bondage and in turn of his liberation; when darkened by the powers of passion it is the cause of bondage, and the cause of liberation when pure of passion and darkness'. (Chas. Johnson's translation, *The Crest Jewel of Wisdom*, p. 20). To put this in other words, the lower mind inflamed by desire and drugged by sloth is the cause of bondage to sin and sorrow; the higher mind illumined by the spirit is the bridge to liberation.

For the planet, man's task is no less than the redemption of its matter both physical and psychical. Man is constantly using the matter of the three worlds for his own purposes thereby educating it to greater responsiveness, sensitivity, and vitality. At the same time he imposes on it his own rhythms, making it more easily responsive to a repetition of the same rhythm. At the physical level he is doing this work with great efficiency. In our material civilization every kind of matter is worked over and used to fill some human need or satisfy some human desire. Man exercises his ingenuity in producing new materials, even to new chemical elements, and in finding new uses for old ones. He has succeeded in releasing energy from the very matter of the planet itself. At the biological level his creative urge — largely exercised for his own advantage it is true — is influencing the course of inheritance. This is being accomplished not only by selective combination of already-existing genetic factors, which has been done for centuries by plant and animal breeders, but by actually producing mutations in the genes themselves. Thus by irradiating seeds and pollen with X-rays and similar radiations several new and useful plant mutations have been produced. This opens up endless possibilities of modifying the course of biological evolution.

At the psychic level, man's influence on matter is at present unconscious though none the less potent. Every thought makes it easier for the matter of the mental plane to respond again to a similar thought. Every emotion similarly by acting on the matter of the astral plane, makes it easier for someone else to feel the same emotions. These effects on the matter of the planes will influence not only succeeding generations, but also succeeding races and even succeeding universes.

Every living creature is as it were a pinhole, a centre through which the divine life streams out into the universe. This life is pure energy, and is neither good nor evil, but neutral. Man has earned the right and has therefore assumed the responsibility of using this life according to his choice. For man has his dual mind and can choose between higher and lower, good and evil, 'ought' and 'ought not'. He can, if he so desires, use the life-stream egocentrically so that it builds a whirlpool with himself as centre and goes round in circles resulting in greater separatedness, isolation, and misery. But the spirit is one and man experiences a constant urge towards unity. Among the creatures he is unique in that he has a higher mind linking him directly with the spiritual monad, and he shows, therefore, the first dawning possibility of a return to the unity from which all was emanated. The higher or abstract mind is a synthesising principle. Synthesis is not the highest aspect of unity, indeed it is only possible because the elements synthesised were already rooted in a deeper intrinsic unity. This deeper unity can be reached through man's intuitive faculty — technically the *buddhic* principle. But even before this is possible, the urge towards unity is still strong. The human level was referred to above as the psycho-social level. It is rightly described thus for as soon as man appears, he organizes and lives in social groups, and sees himself in relation to his group. This social sense is an expression of the higher mind, and today the tendency to organize groups of those with similar interest is surely stronger than ever before.

Every scientific worker, too, has a supreme faith that a unity of principle underlies the diversity of phenomena, and the chief aim of pure science is to find laws embracing these diversities over as wide a field as possible. A. Einstein writing of science has said that it is 'in the rational unification of the manifold that it has its greatest successes'. It is suggested therefore that man's supreme significance in the universe is to act as a unifying agent. This he will achieve most effectively by first achieving unity within his own psyche, leading to unity with God and nature. He would then live freely and spontaneously from his own spiritual centre as an impersonal transmitter of the divine life 'accepting without resistance and giving without reserve' (Rohit Mehta). His infuence will then be on the side of 'the power which makes for righteousness' and he will be living towards greater spirituality, that is towards increasing unity, harmony, and bliss.

CHAPTER 18

VALUES BEYOND

IN THE PREVIOUS pages we have sought to develop the argument that the whole universal evolutionary process is the expression of an unfolding purpose which finds its highest visible expression in man. A long-range survey of the history of human thinking shows that from the earliest times man has speculated upon the ultimate purpose of the universe and it is only locally and sporadically in the long history of philosophic thought that men have doubted that purpose is the very stuff of living.

It is perhaps a sign of the decadence of our present age that it seeks to deny freedom and purposive behaviour to the human spirit. Fashionable patterns of modern thinking and philosophy tend to see individual conduct within a social pattern which is determined wholly by physical heredity and the influence of the environment. Such a view has been repeatedly challenged by thinkers of deeper insight and vision.

Among others, Viscount Samuel has challenged the deterministic and Marxist view of human motive and conduct and the over-emphasis given to sub-conscious motivation in the explanation of human behaviour. The deterministic or negative view of man and society is the result of concentrating upon the abnormal, the infant and the weakling, and upon the deprived and underprivileged in our society. It arises as our psychologists seek to explain the delinquent and the maladjusted individual and the behaviour of immature social groups. Although it is fashionable in some scientific circles to regard science as providing a new and all-comprehensive method which shows the fallacy of any religious and idealistic philosophic vision of the universe and man, there are other trends and tendencies among more thoughtful individuals and groups.

A few years ago a group of scientists in the United States met in conference with clergy and laymen of various Christian denominations and Jewish, Buddhist and Vedanta groups. The purpose of their gathering was to study relations between 'truth and reality' as envisaged by science and religion respectively. A report by R. W. Burke printed in *Science* in 1954 indicates the interest in the very question which is the subject of this transaction. 'Is there a purpose behind the universe?' Scientific and religious thinkers alike expresssed concern about the apparent denial in scientific accounts of human origins of 'spirit', purpose, free will and individual responsibility. Others however indicated that the trouble was more apparent than real and possibly due to verbal confusion. The scientific method is an intensive study of one small field at a time, in temporary isolation, to achieve partial success in understanding, but no observation can be disregarded and the whole structure of knowledge, not excluding religious experience, has ultimately to be interconnected and integrated. As to free will, it was pointed out that causality and freedom are not inherently incompatible and that causation does not necessarily imply compulsion, for our choices are determined by our own motives. There was considerable agreement at this conference that intuition was used by scientists as well as by the formulators of religious truths; far from differentiating them, intuition tended rather to unite the two groups. Although science has undermined some religious dogmas, it is not antagonistic to religion itself. The impact of science might well lead to a gradual discarding of irrelevant concepts and the agreement between different religions upon basic truths, although it is not necessary or even desirable that this should lead to a single world religion.*

The distinction suggested here between science and religion is a useful one, for each has its special field. Scientific and philosophic knowledge should not be confused with the virtues of ethics and religion, such as morality and love, faith and the power of the Spirit. Philosophy and science may be regarded as expressions of ordered knowledge and as such to be employed in creating the conditions necessary to produce human happiness, but the application of these conditions is outside their scope. This is the field of Ethics and Religion, which are purposive in their aim, and involve techniques devised by man for applying knowledge to produce desired ends. Whether the result is immediate happiness and harmony is debatable but in the long run their application will decrease the sum total of human misery in the measure that they 'express the true quality of love that potent universal force recently described as creative altruism'. (Chapter 4). It is in establishing values and principles which will evoke human happiness in an ordered and harmonious society that the ethical and religious viewpoints find their place. The experience of a few individuals who have in all ages been dynamic centres of spiritual inspiration for their fellow men suggests that the development of the spiritual consciousness is a goal worth striving for. All the power and concentration that the individual may possess are needed to bring about this event, and it is possible only if energy is not frittered away along other channels. Like the attainment of any skill, such as that of a ballet dancer, it requires the dedicated purpose of the whole man to achieve the goal decided upon by his spiritual will. The methods for spiritual self-realization are the familiar ones of the age-old story of self-training in yoga and mysticism. Although many may feel such efforts beyond them yet society is under a certain obligation to

---

* See also *Intelligence Came First*, published by the Theosophical Publishing House in America, Wheaton, IL, 1975.

acquaint its members with the goals which are available and considered worth achieving. It is an encouraging sign that books conveying these techniques have been 'best sellers' in recent years.

It has been one of the aims of our survey to show that purpose emerges more and more definitely as we go up the scale of life. Biological species even though determined by the gene-pattern as to their form, move towards goals and express purpose. Man is the agent at the uppermost limit of the life-scale for the working out of a purpose. This is progressively more potent as he develops his conscious mind, acquires bodily skills and techniques, and develops an applied science, itself one vast expression of purpose manifest through human agency. 'The crowning act of creation is the bringing into existence of a being which can in some measure modify the process of which it is a part'. (Prof. R. T. Flewelling at the *World Congress of Faiths* in 1953).

Such a view is, however, not enough for it takes no account of the intent and motive behind man's purposive energies. We must always remember that the goal of humanity is a spiritual and not a material one. It is not the socially delinquent, the neurotic, nor even the multi-millionaires or great industrial barons who evoke our deepest respect and whom we seek most to emulate. The finest type of human being is the one who expresses the greatest wisdom and altruistic compassion and love for humanity. Throughout history these are the ones who have been most venerated as the leading members of the human race. Frequently such men have been of poor physique, with slight and often ailing bodies, and rarely successful in terms of material values. Yet they have embodied ideas and ideals which have changed the world. In venerating the great saints, philanthropists and healers, who have sought to abate a little the sum total of human suffering, even modern society in its judgments puts at the highest point of evolution something which is not material. The scientist looking at the pattern of evolutionary progress tends to look at it purely physically, in terms of the survival of the fittest physical species, the type which has most successfully adapted itself to survival. But from the truly human viewpoint, which includes the deeper aspects of thought and consciousness, it is clear that something different has happened in the human kingdom, which has evolved values other than those of mere survival in competition with others. Our highest impulses have no biological survival value, in fact they are the exact opposite in their consequences. This is a new phenomenon, a new factor in evolution. Such a 'mutation', expressing a purposive spiritual and altruistic drive is something entirely new in the evolutionary pattern of material forms.

The life-force, the *elan vital* as it was termed by Henri Bergson, in the lower kingdoms evolves the forms most biologically suited to survival. It is thus individual in its result, and the evolutionary struggle for existence is set up in terms of the biological organism; but *within* the form at the deeper level there is a more comprehensive expression of purpose. This is Life itself, directive and moving to expression of ever more profound and subtle values which one can only term spiritual and altruistic. The human spirit is itself one with that purposive Life, the dynamic Will of the Cosmic powers, and in understanding all that this implies the insight given by theosophical thinking is required.

In regarding the universe and man as the expression of purposive life, there is no implication of a predetermined end. Human goal images can only be built in terms of past experience, and abstract concepts are distilled or abstracted from such experience. Even our mundane purposes are hardly ever realized as visualized. To try to imagine in formal content the results effected by an ultimate purpose is to put the future in pawn to the past and the present, and it is impossible to visualize, or to describe in words, anything of the nature of final purpose. Goals there must be but they are temporary and will be transcended. Even a cosmic goal is not ultimate if the idea that universe succeeds to universe, be true. Within our human experience, which in its totality is our only guide, each purpose when realized becomes the starting point of a new and wider one. If this process of ever deepening purpose ceases we stagnate and are already half dead. Spirit is always new and vital and in any final sense is transcendent, unimaginable and unpredictable. Humanity, however, occupies a unique place with respect to purpose as to all else.

It must be emphasized once again that for the first time since cosmos emerged from chaos, with the appearance of humanity there comes a creature capable of purposive creative planning and action. As his knowledge of the mechanisms of manifestation increases, and as his creative imagination develops, man becomes increasingly able to foresee and to accomplish novel and hitherto unmanifest ends. These ends are in the cosmic sense only short-term purposes, for man is as yet an apprentice in the art of creation. Many of these purposes are still egocentric and separative but these must in due course become more universalized. Man can become a co-worker with God and with Nature. Indeed he must inevitably become so. It is considerations such as this which move thinkers like Sir Julian Huxley to promulgate the doctrine of 'Transhumanism'. In *New Bottles for New Wine* he says that man is evolving and is the *only* creature that is so doing apart from human intervention. Man *can* become something finer; more intelligent; more far-seeing; more altruistic; less ego-centered; more wise and loving; more nearly self-determining and less the mere sport of circumstances. So the duty devolves on every intelligent and altruistic person to do all in his power to aid in this process of human development.

The present crisis in world affairs could lead to complete chaos and so set back the evolutionary process by many millenia. A world peopled by individualists of increasing knowledge and power carries in itself the certainty of disruption. However in the long run the evolutionary process must continue for it would appear that the increasing emergence of significance, of order and of purpose are intrinsic in the fabric of the cosmos. Man may, if he awakens to his true destiny and his innate divine stature, begin to play an ever-increasing part in the purposive unfoldment of the universe. The choice with which he is faced has been recently stated by Sir Julian Huxley, in *The Destiny of Man*. 'Man is indeed a new and unique *kind* of organism, and has stepped over the threshold of a quite new phase or sector of the evolutionary process. We can call it the human or psycho-social phase. . . . Thus man is not only the most successful dominant type to date, the most advanced product of evolution, but the only type capable of achieving any important advance in time to come. . . . He is the agent of the evolutionary process on this planet.' Having stated our human situation, Huxley goes on to ask, 'what is it man ought to do to fulfil his human destiny?'

In a concluding chapter of *The Destiny of Man*, he suggests that 'Fulfilment' in the widest sense of the word is the goal toward which man's purposive endeavours ought to be directed. Among other essential tasks which such a goal will impose upon mankind: 'it implies the attempt to understand more about man's inner life. How can man resolve psychological conflict; how attain inner peace and spiritual harmony? What is the value of 'mystical' experiences of self-transcendence, and can the techniques of attaining them, like Yoga, be readily communicated and learned? We must follow up all clues to the existence of untapped possibilities like extra-sensory perception. They may prove to be as important and extraordinary as the once unsuspected electrical possibilities of matter.'

Here then, with penetrating insight, are the problems of this new evolutionary era on whose threshold we stand, set out by one of the foremost scientific thinkers of the day. The student of theosophy, accepting as a premise that these questions have an affirmative answer, can meet this challenge as he makes his own experience in this region of man's inner being, and experiments with the techniques and ancient disciplines of Yoga or with the methods which lead to the inner depths of the mystical experience. He may then testify of his own immediate knowledge concerning the reality of the worlds of the psyche and spirit, and tap some of their present little known powers. Since these worlds are realms of spiritual power, harmonious law, and utterly compassionate healing life, this deeper experience will lead men to harmonise their tensions, and solve their problems of 'self-fulfilment' as the interest of more and more thinking poeple is directed to them. The experience of the man who makes this adventure may indeed set others on fire with the flame of the spirit, with its new way of living in interior harmony and brotherly relationship with one's fellow men.

If man can accept this challenge he will become increasingly aligned with the cosmic purpose, with that power that makes for righteousness and can then enter the new era with hope looking forward with confidence to ever expanding horizons.

EPILOGUE

THIS TRANSACTION concludes our series—*This Dynamic Universe* (1951); *This Ordered Universe* (1953); *Man's Expanding Horizon/This Purposeful Universe* (1960)—in which we have sought to re-express in modern idiom the fundamental principles revealed primarily in *The Secret Doctrine*. We have seen the Universe in turn as Energy, as Order, and as Purpose—a newly-formulated Trinity of aspects, within a supremely vital Unity as of one living Being.

Some may consider these Transactions noteworthy as much for what is omitted as for what is said; for we have found little occasion to mention some established theosophical teachings that formed the subject matter of earlier textbooks. It is not that we reject these ideas: they were appropriate in an era concerned with the personal aspects of theosophy, with curiosity about past or future incarnations and the quality of one's personal karma. In these Transactions, however, we have tried to paint a broader canvas. We have by no means ignored man, but we have tried to see the picture from the point of view not of our individual selves but of humanity as a whole.

We have surveyed the inner nature of the Universe from a lofty impersonal viewpoint, and have sought to convey something of the infinite inspiration and grandeur that may be found in this prospect, something of the breathtaking upsurge of spirit it can induce in the receptive mind.

In this stupendous scheme of things, however, man seems almost an incidental intrusion; he is humbled, as by the study of astronomy, taking on an aspect of insignificance in relation to the vastness of the galaxies.

Yet elsewhere, in seeming contradiction, we have taken up an intensely anthropocentric standpoint. We have seen Man at the centre, Man, by whom and for whom all things are created. Here is a new and dangerous doctrine, terribly easy to misunderstand. This is not fallible earthly everyday man as we know him, but Man the Creator, the Heavenly Man, the quintessence of humanity, man yet to be: also the Perfect Man of previous cycles. Perhaps this will be the theme of a new world religion not yet promulgated, though thinly foreshadowed by the teachings of Krishnamurti. But his coldly analytical approach to personality surely needs to be warmed by some doctrine of creative fire, and a joyous sense of togetherness in the tremendous inspiring enterprize still before humanity— to build a heaven here on earth. But such a religion would only be misunderstood by adolescent humanity, and would lead to pride and immodesty. It is rather for adult man, beginning to live from his interior levels, the man not of time but of boundless eternal space.

This apparent contradiction that has emerged from our labours is indeed real and irreconcilable at the mental level. It is like one of the koans of Zen Buddhism, an utter antithesis of viewpoints that becomes a paradox upon insistence that both are true, that indeed they are equally valid statements of one truth. This is the kind of problem that can never be 'solved'; it will not yield to thought alone, nor yet without deep thought; only when the proud mind is exhausted and admits defeat, may the problem disappear by being swallowed up in the fuller understanding that comes in a flash of illumination. But the understanding itself is strictly incommunicable; it can only be translated into hints and analogies.

In this vein let us try to restate the distant goal for humanity. Earlier theosophical writings left the impression that each man, after a multitude of incarnations, would become a sublime individual, and that in some unimaginably distant future it was his destiny to become the Logos of a new system. This idea surely arises from a too literal mental appreciation of a profound truth. It is not each individual man, but in some glorious and mysterious way humanity as a whole that is to become the new Logos. The full meaning of this can only dawn as each man attains Union and enters Nirvana, when 'the dewdrop slips into the shining sea', and 'he is one with life yet lives not', when 'foregoing self, the Universe grows I'. Then, and only then, shall we not merely understand, but become one with, the Purposeful Universe.

We are all in this together: there is no escape. But the *consummation* is deferred until *all* are ready. For the purpose of the Universe is to create a new God — and we are that God.